Murray-Darling Mysteries

NATURE'S UNIQUE STRATEGIES FOR SURVIVAL IN THE MURRAY AND DARLING RIVERS IN AUSTRALIA

ANNE JENSEN

Ordering Information:

Prime Seven Media
518 Landmann St.
Tomah City, WI 54660

Printed in the United States of America

Dedication

This book is dedicated to the late Associate Professor Keith Walker, long-time Murray River Champion, with thanks for his vision and passion for expanding our knowledge of these vital river systems. It was Keith's ambition to complete a natural history of the Murray River Valley, to share his knowledge and love of the river ecosystem with a wide audience.

Keith was a passionate advocate for understanding the ecological rhythms and connections of Murray-Darling Rivers, with special attention on overlooked species, like mussels, black box and lignum. He was pivotal in guiding and shaping decades of research work on Murray River ecology, encouraging students to study multiple aspects of natural river and floodplain ecosystems. Keith identified the importance of river flows in sustaining river ecosystems and described flow as the 'maestro' conducting the orchestra of river life.

In his memory, this book aims to to complete his vision, building on the shared knowledge and experience of many of Keith's colleagues, by sharing easy-to-read stories of the mysteries of how natural Murray-Darling River ecosystems work and how they have changed and adapted since European settlement.

"Let's have fun doing good science. Let's not forget the wonder of it all, and let's make sure the people who follow us can also see the Murray River the way it was, in the beginning."

Associate Professor Keith Walker, quote from Hillary Jolly lecture, Australian Society of Limnology 1993

Prologue

The Murray and Darling-Baarka Rivers in south-eastern Australia are full of intriguing mysteries. How have their unique flora and fauna developed clever strategies to survive in widely varying conditions which swing between droughts and floods?

In one of the world's oldest and slowest river systems, Murray-Darling rivers have evolved their courses and floodplains over millions of years on the planet's oldest and driest continent. The Murray River meanders more than 2300 km across a semi-arid ancient landscape from the mountains to the sea. The Darling flows more than 2800 km through even drier landscapes to its junction with the Murray River.

While their history stretches back 60 million years through many different pathways, the Murray and Darling Rivers have only followed their current courses for the last 7-8000 years. The characteristic plants and animals, including river red gums, frogs, fish, waterbirds and turtles, are dependent on the pattern of river floods, droughts, high and low flows. Majestic river red gum trees lining the rivers are several hundred years old, germinated long before European settlement in Australia in the late 1780s brought big changes to the rivers and the way they flow.

The river plants and animals have developed harmonies with the highly variable conditions which allow them to thrive and survive. Unravelling the mysteries and understanding the natural cycles and systems of the Murray-Darling Rivers will help to manage the river systems wisely for the future.

Dramatic sandstone cliffs define the edges of the narrow Murray Gorge downstream of Morgan in the South Australian Riverland

Foreword

The Murray-Darling Basin is a magnificent river system. The Darling and the Murray Rivers start their journeys west from high up in the mountains of the Great Dividing Range. Their tributaries, like outspread fingers, flow west to join the wrists that are the Darling River and Murray River. This extensive river basin provides habitats for an amazing range of different plants and animals. The ecological health of the rivers depends on their connectedness, from the algae, plants, other microbes, crustaceans and insects, which are at the bottom of the food chain to the turtles, frogs, native fish and waterbirds at the top. None of this magic can happen without the essential natural rhythm of water flow. As the late Keith Walker would say, flow is the 'maestro' in the river orchestra. Flow connects the biological world and links the river channels to the floodplains and everything in between. Our river communities, the people in towns, First Nations' communities, fishers, graziers and irrigators all depend on the Basin's river flows and good quality water.

Our harnessing of Murray-Darling river water, with dams and diversions of water, primarily for irrigation, has changed these mighty rivers for ever. There are consequences everywhere: up and down the full length of the Murray-Darling Basin. Dead and dying floodplain eucalypts, declining waterbird, and frog communities, native fish kills, more algal blooms and rising salinity signal rivers in poor shape. However, natural floods and environmental water can quickly resuscitate them. As this book illustrates so well, there is still much to wonder and enjoy about the rivers and wetlands of the Murray-Darling Basin and it is incumbent on all of us to restore these rivers.

This book is a wonderful tribute to Keith Walker, a lovely man who dedicated his life's work to the Murray River. He was an inspiration to me with his great passion and enduring optimism. He was a scientific pioneer in unlocking the biological rhythms of the rivers. Please read this book and build your understanding and appreciation of this great river system. Most of all, appreciate the overriding importance of water and flow to its future and all the communities and other organisms that depend on it. We owe it to future generations to protect what is left of these special river systems.

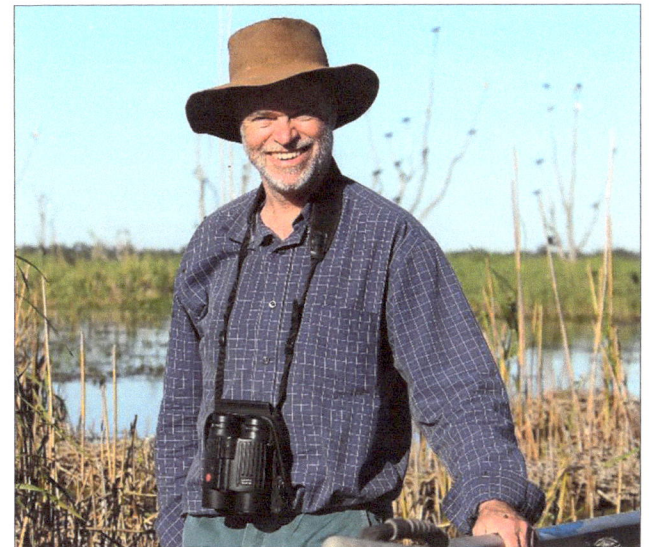

Professor Richard Kingsford, Professor of Environmental Science, Director of Centre for Ecosystem Science, School of Biological, Earth and Environmental Sciences, University of New South Wales, Sydney, Australia

Acknowledgements

The stories told in this book have been gathered over long years of collaboration on Murray River research and management with Keith Walker and many colleagues. In the 1980s a series of seminars called 'Murray Mysteries' provided expert presentations about Murray Valley science for river tourism operators. In 2009 the 'Water is Life' project collated science-based stories by University of Adelaide researchers for an interactive Murray River trail at Adelaide Zoo.

Special thanks for shared wisdom to Emeritus Professor Jane James (geology), Associate Professor David Paton (birds and the Coorong), Professor Richard Kingsford (birds, red gum health and the impacts of changing flows), Mr Paul Otto (birds) and Dr Scotte Wedderburn (fish). The stories about mussels, crayfish and yabbies, snails, fish and the impacts of weirs on rivers and ecosystems all came from Associate Professor Keith Walker. Thanks to Dr Fiona Paton and Dr Tom Hunt for sharing their beautiful bird photos.

And a special thank you to Jan Walker for her careful review of the material to make sure the stories were told in a way that Keith would have approved.

River red gums line the Murray River mainstream along the broad river valley near Berri in the South Australian Riverland

Table of Contents

Introducing the Murray-Darling Basin and the Murray and Darling Rivers

The Murray-Darling Basin is the combined area that drains water into the Murray and Darling Rivers. It covers one-seventh of Australia's land mass, more than one million km^2, an area as big as France and Spain combined. The headwaters lie in the Great Dividing Range of eastern Australia, with the Darling River system rising in Queensland and New South Wales, and the Murray River and its tributaries rising further south in Victoria and New South Wales.

The headwaters of the Murray River start in the Snowy Mountains in south-eastern Australia, near Australia's highest point of Mt Kosciusko. The mouth of the Murray River is over 2,300 km away, on the south coast of South Australia near the river port of Goolwa. In contrast, the Darling River wanders more than 2,800 km from southern Queensland across semi-arid flat plains to join the Murray River at Wentworth in western New South Wales.

While it is Australia's most important river system, a whole year's flow from the Murray-Darling system is less than one day's flow from the Amazon River in South America. By far the greatest volumes of water flow from the Murray River and its main tributaries in the Southern Basin. Rainfall and snow melt in this catchment are the main sources of flows to the downstream end of the system in South Australia, naturally peaking in mid to late spring.

Flows from the Darling River are highly variable, contributing on average only 10% of flows into the Murray River. These flows tend to arise from summer monsoons, arriving in time to extend peak flows in the Murray River.

The extent of the Murray-Darling Basin, covering the states of Queensland, New South Wales, Victoria and South Australia, and including the Australian Capital Territory around Canberra. The Murray River catchment, arising in the Snowy Mountains, contributes the highest volumes. It forms the border between Victoria and New South Wales
(Source: © Murray-Darling Basin Authority 2020)

In spite of its low flow, the Murray-Darling Basin is Australia's most productive river system, supporting irrigation and agricultural production worth over $15 billion every year. More than 85% of Murray-Darling water extracted is used in the upstream states of New South Wales and Victoria for agricultural production. South Australia receives 8.6% of water, Queensland 5% and the Australian Capital Territory 1%.

The Murray-Darling Basin supplies drinking water to nearly 3 million Australians, including 75% of South Australia's population in the driest state in Australia, which is reliant on the Basin as critical back-up for water supply. On average, about 40% of South Australia's water comes from the Murray River, but in drought years this increases to over 90%. Water supply for 1.1 million people in metropolitan Adelaide is about 1% of water taken from the Basin.

The rivers and their natural ecosystems are full of fascinating plants and animals, with specially-developed strategies to cope with the unreliable climate and extreme conditions of the 'land of droughts and flooding rains'[1]. These plants and animals are adapted to survive the dry times and make the most of the wet times. They have a variety of responses to 'boom and bust' conditions so that they can survive and thrive on the floodplains, river banks, and in the rivers themselves. Healthy ecosystems mean healthy rivers and clean reliable water supplies.

Since European settlement, more and more water has been extracted from the rivers of the Murray-Darling Basin. As the amount of water in rivers has reduced and flow patterns have changed, native plants and animals have diminished in numbers and some struggle to survive, even to the point of local extinctions. River ecosystems have lost their resilience, particularly their ability to cope with drought, because the chances to replenish water stores in wet times have been reduced. For plants and animals in the Murray River and Darling River ecosystems, water is critical to their well-being and survival, and to sustain healthy river ecosystems. Equally, river communities, irrigation industries and South Australian cities and towns rely on **healthy working rivers** for their livelihoods.

The key message is that 'Water is life', for everything dependent on healthy river systems: humans, plants and animals. To maintain healthy working rivers in the Basin, understanding how river species have adapted to the variable water regime of the natural Murray-Darling Basin environment is paramount. A healthy river system needs sufficient water to deliver essential ecosystem services which will support healthy human communities and a healthy economy.

Key Points

- the plants and animals of the Murray-Darling Basin are adapted to the 'boom and bust' conditions of highly variable flows, alternating between floods and droughts
- In spite of low volumes of water, the Murray-Darling Basin supports most of Australia's irrigated produce
- since European settlement, increasing amounts of water have been extracted, affecting the health of river systems and decreasing their ability to support river- dependent communities and industries

Dramatic eroding sandstone formations catch the golden afternoon sun at Headings Cliffs near Renmark, South Australia

---

[1] An often-quoted phrase from a well-known poem 'I love a sunburnt country' by Dorothea Mackellar

Riverland artist Garry Duncan has captured the essence of the iconic river habitats of Murray-Darling rivers in his vibrant painting 'Southern Wetlands'. In his unique style, he portrays the magic of a typical river scene with its reflections and ripples.
Using linked panels, he illustrates the complexity and varied species of the river ecosystems.

The left panel depicts 10 key waterbird species, the centre panel includes 7 aquatic species such as water rats, fish, yabbies and frogs, and the right panel has 11 more waterbirds which rely on river habitats.

The composition cleverly suggests how they are all interlinked and inter-dependent on the river systems.

The original painting is owned by the author, and artist Garry Duncan has given permission for its reproduction here.

Aboriginal History of Murray-Darling Rivers

Aboriginal people arrived in Australia about 65,000 years ago and had reached the Murray-Darling Rivers by 40,000 years ago. There were many Aboriginal settlements along the rivers, with reliable sources of food and water. Their rich cultural history is found in campsites, fireplaces, flints and stone tools, middens, canoe and shield scars, rock engravings and their dreaming stories.

One of the richest archaeological sites at Lake Mungo indicates that Aboriginal people had settled around the Willandra Lakes at the end of the Lachlan River catchment around 40,000 years ago, gathering perch, shellfish, lizards and emu eggs from the freshwater lakes system and surrounding lands. During wet cycles the lakes filled with fresh water and supported substantial fish and mussel populations. Fish ear-bones and mussel shells have been found in middens and hearths, providing dates for those wet cycles. The last Ice Age occurred 20,000 years ago, when there was snow on mountains and hills across the Basin, less rain and lower water levels. The climate became warmer by 14,00 years ago, and the Willandra Lakes dried out.

The freshwater Lakes Alexandrina and Albert first formed at the end of the Murray River 7-8000 years ago, when the sea reached its current level. By the time European settlers arrived, the Murray Valley was supporting more Aboriginal groups than most other places in Australia.

European settlement began in the late 1700s, when squatters took over river banks and plains for their sheep, cattle and crops. Aborigines were displaced from their traditional lands and many died from influenza and small pox. From the late-1800s, most were forced to leave their lands and move to mission stations.

Today Aboriginal Traditional Owners have formed their own communities and farms, and many now live in riverside towns to keep their connection with rivers. There are 46 First Nations across the Murray-Darling Basin. They are gaining increasing recognition and engagement on river management issues, including securing cultural river flows, via representation from Northern Basin Aboriginal Nations (NBAN) and Murray Lower Darling Rivers Indigenous Nations (MLDRIN).

Lake Mungo archaeological site in western New South Wales covers a group of ephemeral lakes at the end of Willandra Creek, a remnant of the Lachlan River catchment: eroding lunette on the east shore of Lake Mungo known as the Walls of China; stone tools and chips; charred clays of a hearth dated to 28,000 BP

Key Points

- Aboriginal people have been living in the Murray-Darling Basin for more than 40,000 years
- Aboriginal culture is highly interwoven with rivers, their flow patterns and ecosystems
- Dreaming stories tell of the creation of rivers and the plants and animals
- when European settlers arrived, there were many Aboriginal settlements along the rivers
- Aboriginal links to their traditional lands and rivers were severely disrupted by European settlement
- Traditional Owners are now re-establishing links and claims for cultural flows to maintain their traditional activities

Rock engravings on the cliff face at Ngaut Ngaut in the Lower Murray Valley

Aboriginal Names for Murray River and Darling River

The Lower Murray River below the Darling River junction flows through the lands of various Indigenous groups, including Ngintait, Erawirung, Ngawait, Ngaiawang, Nganguruku, Ngaraita, Portaulun, Jarildekald, Warki and Ngarrindjeri at the end of the river system. Aboriginal names for the Murray River change between Indigenous groups and include: Ingalta, Moorundie, Goodwarra, Parriang – Kaperre, Tongwillum, Yoolooarra.

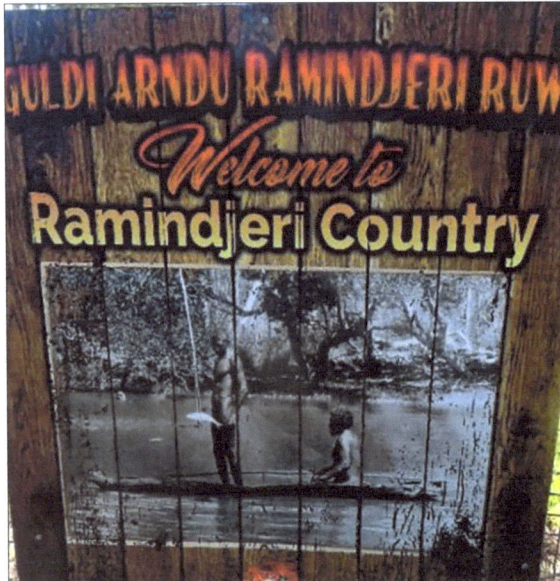

Ramindjeri country includes Victor Harbor, the Inman Valley and Encounter Bay, all west of the Murray Mouth

Variations of 'Moorundie' are frequently found at locations in the Lower Murray Valley.

The Coorong, Lower Lakes and Murray Mouth region has been the homeland of Ngarrindjeri people for thousands of years, and their culture and traditions remain vibrant. There are four clans: Ramindjeri from Encounter Bay, Tangani from the Coorong, Yaraldi from east of the Lakes and Lower Murray, and Warki from western Lake Alexandrina. The Ngarrindjeri Regional Authority was formed in 2007 and is overseeing the development and implementation of the Ngarrindjeri Nation Yarluwar-Ruwe Plan (*Sea Country Plan*).

The Barkindji people identify with the Darling River (Baarka) and Barkindji culture revolves around the Baarka River. Their storyline lies within the river and its dependent ecosystems, with their flow patterns, plants and animals. The Barkindji identify strongly with the river, its health and its flows.

The Ngurunderi Dreaming Story

The creation of the Murray River is told in the *Ngurunderi Dreaming Story* of the Ngarrindjeri People.

Ngurunderi was an important law-giver and the main shaper of the distinctive landscape of the Murray River. In the dreaming story, *Ngurunderi* travelled down the Murray River in a bark canoe, in search of his two wives who had run away from him. At that time the river was only a small stream below the junction with the Darling River.

Large Aboriginal canoe scar on a giant river red gum by the Murray River at Chowilla

The Ngurunderi dreaming as told by Albert Karloan, one of the last initiated Ngarrindjeri men, to anthropologist Ronald Berndt (1940). From the Ngurunderi Gallery 'The Culture of the Ngarrindjeri People', South Australian Museum, Adelaide.

A giant river cod fish (*Ponde*) swam ahead of *Ngurunderi*, widening the river with sweeps of its tail. *Ponde*'s tail also made swamps and cliffs along the way. *Ngurunderi* chased the fish, trying to spear it from his canoe. Near Murray Bridge he threw a spear, but it missed and was changed into Long Island (*Lenteilin*). At Tailem Bend (*Tagalang*) he threw another spear; the giant fish surged ahead and created a long straight stretch in the river. At last, with the help of *Nepele* (the brother of *Ngurunderi*'s wives), *Ponde* was speared after it had left the Murray River and had swum into Lake Alexandrina. *Ngurunderi* divided the fish with his stone knife and created a new species of fish from each piece, including golden perch, silver perch and bony bream. He threw the last piece back into the lake and said 'go on being *Ponde*'.

Meanwhile, *Ngurunderi*'s two wives had made camp. On their campfire they were cooking bony bream, a fish forbidden to Ngarrindjeri women. *Ngurunderi* smelt the fish cooking and knew his wives were close. He abandoned his camp and came after them. His huts became two hills and his bark canoe became the Milky Way. Hearing *Ngurunderi* coming, his wives just had time to build a raft of reeds and grass-trees and to escape across Lake Albert. On the other side their raft turned back into the reeds and grass-trees. The women hurried south.

Ngurunderi followed his wives as far south as Kingston. Here he met a great sorcerer, *Parampari*. The two men fought, using weapons and magic powers, until eventually *Ngurunderi* won. He burnt *Parampari*'s body in a huge fire, symbolised by granite boulders on the beach. Then he turned north along the Coorong beach. Here he camped several times, digging soaks in the sand for fresh water, and fishing in the Coorong lagoon.

Ngurunderi made his way across the Murray Mouth and along the Encounter Bay coast towards Victor Harbor. He made a fishing ground at Middleton by throwing a huge tree into the sea to make a seaweed bed. Here he hunted and killed a seal and its dying gasps can still be heard among the rocks. At Port Elliot he camped and fished again, without seeing a sign of his wives. He became angry and threw his spears into the sea at Victor Harbor, creating the islands there. Finally, after resting in a giant granite shade-shelter on Granite Island (*Kaike*), *Ngurunderi* heard his wives laughing and playing in the water near Kings Beach. He hurled his club to the ground, creating the Bluff (*Longkuwar*), and strode after them.

His wives fled along the beach in terror until they reached Cape Jervis. At this time Kangaroo Island (*Karta* - the land of the dead) was still connected to the mainland (18,000 years ago), and the two women began to hurry across to it. When *Ngurunderi* arrived at Cape Jervis, seeing his wives still fleeing from him, he called out in a voice of thunder for the waters to rise. The women were swept from their path by huge waves and were soon drowned. They became the rocky Pages Islands.

Ngurunderi knew it was time for him to enter the spirit world. He crossed to Kangaroo Island and travelled to its western end. After first throwing his spears into the sea, he dived in, before rising to become a star in the Milky Way. Before *Ngurunderi* left the earth to live in the Milky Way, he told the people 'I am going first, you will come after me.'

The Murray River Dreaming story says the Bluff or Longkuwar at Encounter Bay was created when Ngurunderi hurled his club into the ground. Wright Island was one of the islands created when he threw his spears into the sea

Murray River through the Millenia

The Murray-Darling Basin and its multiple rivers have a history stretching back 60 million years. The rivers evolved along with Australia's unique flora and fauna in wetter times, with sea levels sometimes higher, sometimes lower than present. The course of the rivers has changed and re-formed many times.

When sea levels were higher, the Murravian Gulf extended inland as far as modern-day Swan Hill for more than 40 million years. The shallow, sheltered waters of the gulf supported prolific shellfish, including cockles, oysters, sea urchins and starfish. Large areas of limestone formed on the bed of the Murravian Gulf and, as the Basin subsided over millions of years, very large numbers of marine fossils formed in the limestone.

Fossils in remnant limestone layer at Shell Hill

From 6-2 million years ago, the fore-runner of the Murray River flowed in a shallow valley, and large lakes and wetlands developed. About 2 million years ago, tectonic uplift dammed the Murray River south of Swan Reach in South Australia to form *Lake Bungunnia*, with water backed up to Swan Hill in Victoria. Lake Bungunnia existed for more than a million years, with abundant flora and fauna.

About 700,000 years ago, the tectonic dam downstream of Swan Reach was breached and Lake Bungunnia drained rapidly, with the River scouring down through limestone sediments on the lake bed.

Between 30--25,000 years ago the Murray River established its current drainage pattern. The last Ice Age (25--16,000 years ago) produced a cool and wet phase, with temperatures 8°C cooler than today. There were glaciers in the Snowy Mountains, and sea levels were 130-160 metres lower than today. The Murray River flowed off the edge of the continental shelf south of Kangaroo Island, 200 km south of today's coastline, forming deep sub-marine canyons. The lowered sea level increased the gradient and flow velocity, causing the Murray River to cut deeply into the limestone deposits from the Murravian Gulf more than 30 million years previously. This process produced towering limestone cliffs and created the Murray Gorge between Overland Corner and Mannum.

In the period 16--13,000 years ago, the climate warmed, the glaciers in the Snowy Mountains melted and the rivers became narrower and more sinuous. Trees returned to the landscape in the western part of the Basin and on the riverine plains.

Key Points

- the Murray-Darling Basin is 60 million years old but the modern Murray River only started to form 6 million years ago
- the shallow Murravian Gulf formed 50 million years ago
- layers of limestone full of marine fossils were laid down in the Gulf
- 2 million years ago Lake Bungunnia formed, from Swan Reach upstream to Swan Hill
- 700,000 years ago Lake Bungunnia drained and the river cut down through the limestone sediments
- Aboriginal people have been living near rivers and wetlands across the Basin for the last 40,000 years
- the modern river course developed 25-30,000 years ago
- 16-25,000 years ago, sea levels were much lower and the river cut deeply into limestone sediments, forming the Murray Gorge
- current climatic conditions appeared 8-10,000 years ago
- river channels changed near Echuca 8000 years ago and current sea level was reached 7-8000 years ago, so the modern Murray River course is 7-8000 years old

For the past 10,000 years the climate of the Murray Basin has been similar to today. The Cadell Fault upstream of Echuca in the mid-Murray blocked the flow of the river about 8000 years ago, causing flows to spill seasonally across the floodplains in the large triangular Barmah-Millewa floodplain, supporting the world's largest river red gum forest.

The Murray and Edward Rivers found new courses around the barrier of the raised Cadell Fault. The main path of the Murray turned south to flow past Echuca. A second major stream formed into the Edwards River flowing north and then west, breaking into multiple braided streams through the Wakool region. The streams met up again as the Murray River downstream of Swan Hill.

The Lower Murray River below the Darling River junction has the characteristics of an old river. No tributaries enter the Murray River below that point. The river has a very low gradient and flows very slowly, with extensive meanders, cut-off oxbows, anabranches and extensive floodplains. This broad valley form predominates from the Darling River junction to Overland Corner in the South Australian Riverland.

The Murray River meanders across the broad floodplain upstream of Loxton in South Australia

From Overland Corner to Mannum, the river changes character abruptly, confined to the narrow Murray Gorge which is only 1-2 km wide and bordered by distinctive sandstone cliffs. The main channel meanders back and forth between the cliffs, with pockets of floodplain and sandy river beaches on the inside of meander bends.

Over its long geological history, as sea levels rose and fell, the Lower Murray Valley deposited alluvium on the valley floor to an average depth of 30-40 metres. During periods when sea levels fell, the river channel was carved out of the deep silt, leaving the current river without a solid bed, particularly downstream of Murray Bridge.

The river channel empties into two large terminal lakes, and the estuary features extensive sand barrier islands, meandering channels and mud flats. About 7-8000 years ago the sea reached its modern level and the river estuary has been in its present complex configuration of channels and barrier islands for about the same period. The Murray Mouth lies on the high-energy, dynamic Southern Ocean coast. The Mouth has been around its present position for at least 7,000 years but it shifts continually back and forth over about 6 km, driven by river outflows versus ocean and sand inflows from tides and storms.

Sandstone cliffs of the narrow Murray Gorge near Woolpunda in South Australia

How Water Flows in the Murray and Darling Rivers

The Murray-Darling Basin, which covers more than one-seventh of Australia, is fed by 27 rivers and their tributaries, and crosses four states and one territory. Water falling on the Great Dividing Range along Australia's east coast is channelled inland to meander through vast semi-arid plains. Water reaching the sea at the Murray Mouth in South Australia travels over 3600 km from the headwaters of the Darling River in Queensland or 2500 km from the headwaters of the Murray River in the Snowy Mountains.

Most of the water in the Basin comes from the Murray River catchment, with three-quarters of the run-off in the Murray-Darling Basin coming from the small area of the Snowy Mountains in south-eastern Australia. Additional inflows to the Murray River come from a series of tributaries also rising in the Great Dividing Range in Victoria, including the Kiewa, Ovens, Goulburn, Loddon and Campaspe Rivers. In New South Wales significant flows reach the Murray River from the Murrumbidgee River catchment. The Lachlan River rarely flows through to the Murrumbidgee, spreading out into the Great Cumbung Swamp.

The Darling River catchment channels flows from the northern basin into the Murray River at Wentworth in western New South Wales. Its catchment includes the Condamine, Balonne and Barwon Rivers which arise in Queensland and the Macquarie, Gwydir and Namoi Rivers flowing through New South Wales. The Macquarie River flows through the internationally important Macquarie Marshes wetland on its way to the Barwon River which flows into the Darling River.

Flows in the Darling River vary widely in response to monsoonal rains, from minimum low flows to floods. The Darling River fills the important Menindee Lakes system on its way to the Murray River and supplies key regional towns along its route.

Water flowing from the Darling River resembles iced coffee due to the heavy load of clay particles carried from its floodplain. Water flowing from the Murray River usually resembles iced tea due to leaf tannins, with an increasing green tinge from algae during low flows. From Wentworth and downstream into South Australia and to the Murray Mouth, water from the two rivers is mixed and the Murray River is usually dominated by the milky brown colour of the

At Wentworth in western New South Wales, the waters of the Murray River (foreground) and Darling River (right centre) merge and mix as they flow over the weir at Lock 10 (top left).

Key Points

- 27 rivers channel water into the Murray-Darling Basin across Queensland, New South Wales and Victoria
- water flows over 2500 km to reach the sea via the Murray River and 2800 km via the Darling River to the Murray Mouth in South Australia
- the headwaters of the Murray-Darling Basin rivers are swift-flowing streams in steep valleys
- across the lowland plains, the rivers become slow-flowing, meandering in broad valleys because of low gradients
- more than half of the water in the Murray-Darling Basin flows through the Murray River, fed by run-off from the Snowy Mountains in the south-eastern corner of Australia
- the Murrumbidgee, Kiewa, Ovens, Loddon, Campaspe and Goulburn Rivers contribute nearly 40% of flows in the southern Basin
- the Darling River catchment in the northern Basin has highly variable flows, dependent on monsoon rains which usually peak in summer

Darling clays. There are no significant tributaries to the Lower Murray within South Australia, with minor seasonal inflows from the Mt Lofty Ranges reaching the Lower Lakes via the Marne, Angas and Bremer Rivers.

Time for Water to Travel from Source to Sea

The longest distance within the Murray-Darling Basin is from southern Queensland in the Darling River catchment to the Murray-Darling junction at Wentworth and on to the Murray Mouth, a distance of over 3600 km. In the headwaters of both Murray and Darling Rivers, tributary streams flow swiftly in steep valleys, with pools and riffles supporting local fish and their food sources. Once they reach the flat inland plains, the rivers meander lazily because of the low gradient and slow flows, with only 1400 m height difference over more than 2300 km for the Murray River. A drop of water takes eight weeks to travel 2200 km down the highly regulated Murray River, from the Hume Dam near Albury to the sea. Travel times down the highly variable and less regulated Darling River are much more difficult to estimate. From the Murray-Darling junction at Wentworth, in the fully-regulated Murray River, a drop of water takes about four weeks to travel a meandering 827 km to the sea, with only 75 cm fall over the last 270 km from Blanchetown to the river mouth.

How the Rivers Flowed Naturally

Lazy river meanders in the Lower Murray Valley near Weir & Lock No 9 at Kulnine

The Murray River had a natural seasonal cycle of high flows in spring and low flows in late summer and autumn. In very dry years, the river dried back to lower levels but recent research into fish life cycles suggests the main channels never dried up and native fish were able to survive through low flows. Similarly, the Darling River had low flows in dry years, allowing fish populations to survive droughts.

In wet years, the rivers expanded over their floodplains, triggering major breeding and regeneration events in native plants, birds, fish and frogs. In dry years, flows reduced, with rivers becoming too shallow for paddle-steamers but still had enough water flowing to sustain large native fish like Murray cod.

Much of the Murray-Darling Basin contains marine sediments, so saline groundwater entering the rivers during low river flows could make the water quite salty. In 1830 near the Darling junction, the explorer Captain Charles Sturt found river water was too salty to drink.

Where the Murray River reaches the Sea

The Murray River flows into Lake Alexandrina and Lake Albert, two large terminal lakes, often called the Lower Lakes. At settlement, the Lower Lakes were predominantly fresh, with occasional intrusions of seawater during very low river flows when tidal inflows were greater than river outflows. In the early days of settlement in the 1840s, the Lower Lakes provided fresh water to towns and enterprises on their shores, including a wool-washing plant near Narrung relying on freshwater. However, the impact of water being withdrawn upstream was already having an effect by the 1870s and the wool-washing plant was forced to close because the water became too salty.

Dynamic Murray Mouth

The natural estuary of the river, where river water mixed with sea water, lay between Point Sturt at the end of Lake Alexandrina and the Mouth. The high energy coast pushed sand in through the Mouth and the river pushed it out again during high flows. This process produced a chain of barrier islands made of mud and sand, and the river had to meander its way around them to reach the sea. The river mouth was naturally highly mobile, migrating along the coast over a range of up to 6 km over the last 3000 years, moving back and forth over 1.6 km in the past 190 years. The original Murray estuary was an important nursery habitat for fish and waterbirds.

Water Flowing out to Sea is not Wasted

Water flowing through the river systems and out to sea is used over and over again, transporting nutrients and biota along the way, as well as providing water to towns and communities. River water supports towns, businesses, agriculture, tourism, regional communities and the national economy. It keeps rivers flowing to support native fish, waterbirds, frogs and vegetation, and flows all the way to the Lower Lakes, Coorong and the sea. It refreshes water in the Coorong, preventing build-up of salinities and toxins above the tolerance of the fish and waterbirds which rely on its food sources. Water flowing out into the Southern Ocean carries salt out of the Basin and also supplies food and nutrients for the commercially important crayfish nursery off-shore. Flows to the end of the system and out to sea maintain the health of both the river systems and ocean systems, as well as for river communities along the way. This water absolutely is not wasted.

Aerial view of Murray River entering Lake Alexandrina (top right) and the main river channel flowing west from Point Sturt around Hindmarsh Island to reach the Murray Mouth. (Base map: NatureMaps SA)

Floodplains Need to be Flooded

The floodplain is the broad, relatively flat expanse of the river valley on either side of the mainstream where floodwaters can spread out. It is formed by a succession of flood flows spilling out over this area, working and re-working the sediments into a complex of terraces, hollows, wetlands, billabongs, backwaters and anabranches. The floodplain provides room for high flood flows to spill out of the mainstream, pond over a large area and slowly drain back to the river, leaving behind sediments and nutrients which are taken up into food chains, and returning microscopic plants and animals to the mainstream as floodwaters recede.

Floodplains are a flood buffer, a water filter, a nursery and the centre of biological life in river systems. Key steps in breeding and regeneration cycles occur on the floodplain, with most species dependent on flooding and floodplain habitat for their survival. Floodplains play a key role in maintaining water quality in river systems. They provide a means of freshening and oxygenating wetlands and backwaters, exchange old water for new, dilute salts and nutrients, and create healthy habitats for birds, fish and plants to grow and survive.

Floodplains are created and renewed by flooding. They are ever-changing places. They require inundation with water to replace nutrients and to flush out accumulated salts and de-oxygenated water. Floodwaters spilling onto floodplains bring new supplies of seeds, microscopic plants and animals, nutrients, soils and oxygen, triggering a new generation of plants and animals to sustain the river valleys and return new life to the rivers.

Rivers and their floodplains need to be managed together. Together they make up a single ecosystem that is interlocked at all levels. The relationship between floodplains and rivers has been compared to the relationship between bark and trees. If rivers are not allowed to overflow onto their floodplains and receive enriched waters back as the floods recede, rivers will slowly die, like a ring-barked tree.

Water spreading over floodplains triggers a flush of growth of microscopic plants and animals which support breeding of birds, animals and fish

> ## Key Points
>
> - floodplains are an essential part of river ecosystems
> - rivers flooding onto their floodplains reduces damage from flooding, slowing down flows, lowering flood levels and allowing sediment to drop out
> - the process of water spilling out onto floodplains and later returning to rivers triggers breeding and germination in native plants and animals
> - nutrients are extracted from the water and taken up in the food chain
> - accumulated salt and nutrients are washed off the floodplains
> - salty groundwater under the floodplain is refreshed
> - water is stored in floodplain soils to support plants and animals through dry times

Benefits of Floods

Floods threaten people's homes and livelihoods, so they are seen as bad. Newspaper reports talk about people battling against the forces of nature, defending their homes and their towns. However, floods are essential to maintain river ecosystems and provide benefits to river communities and industries, such as clean water, fish and tourism assets.

Local communities benefit from the richer, more productive soils of the floodplain, healthy trees lining river banks and improved fishing during extended periods between the infrequent inconveniences of floods. It is unfortunate that many towns and farms have been built low on the floodplain where they are at greater risk of being flooded occasionally. These towns now argue that they should be protected from floods and rivers should be prevented from spilling onto floodplains, but the benefits of floods on floodplains need to be recognised.

Floods maintain river health and the land and water resources that both farmers and communities rely on: the soil, water, trees, birds, fish, insects and river landscapes. The floods support the life cycles which provide fish and yabbies for amateur fishermen, recreation and tourism opportunities, and shady trees along river edges. If fishermen want to catch yabbies for Christmas, there needs to be a spring flood in October-November!

Broad Murray Valley near Moorook in South Australia, in normal flows (left) and flooded in 1993 (right)

Timing of Murray and Darling Floods

High flows in the Murray River naturally occur in spring, from September to early December, fed by winter and spring rains and snow in the Snowy Mountains catchment. Unpredictable floods from the Darling River are more likely to occur in summer, mostly in December-January, as a result of tropical monsoonal rains.

River floods are described in terms of the chance of flooding each year. Before dams and weirs, a small flushing flow within the main channel occurred nearly every year. A flow of 35,000 megalitres per day (ML/d) which spilled onto the floodplains in the Lower Murray Valley had a chance of occurring 1 in 2-3 years. Bigger floods of up to 100,000 ML/d had a chance of occurring 1 in 5 years.

With dams and weirs in place, the chances of floods have been more than halved, with almost all of the small floods captured in dams and no longer reaching the floodplains. Flows similar to the 1993-94 flood peak of 126,000 ML/d now have a chance of occurring only 1 in 10-12 years with flow volumes about half of what would have flowed naturally without the upstream dams.

Murray River channel near Woolpunda in 2016 flood, with rising levels already 1 m higher than normal levels in October (left) and overflowing onto the floodplain in December (right)

A Wealth of Habitats on Floodplains

Floodplains support of a wealth of diverse habitats, ranging from billabongs, horseshoe lagoons and waterholes, to red sand hills, claypans and lunettes. The plants and animals found on floodplains reflect how often those habitats have water. Plants range from wetland species to drought-tolerant trees and shrubs like saltbush. The drier, higher areas of floodplains support pockets of dryland vegetation, including native pines and spinifex. A whole range of wetland types are found, from permanent pools and creeks to temporary swamps filled by floods and ephemeral areas which only become wetlands in the highest of floods or after heavy rain.

Snags and Litter are very Important

Snags (fallen trees in the water) and natural litter like bark, twigs and leaves form very important habitats on floodplains and in the water. Branches and trees which fall into permanent water provide important underwater habitat. Algae forming on their surfaces provide food for grazing animals and fish. Hollows in the branches provide protected underwater nests for fish. River red gum wood persists for decades underwater, ensuring the long-lived Murray cod of a secure home.

Litter on the floodplain includes fallen branches, twigs, bark, leaves and dead plants. This provides mulch for the soil, hiding places for small creatures and prevents evaporation from the ground surface. A specialised food chain operates within the litter, breaking it down into soil and recycling the nutrients. Eucalypt leaves falling into the water provide food for specially-adapted grazing insects and other *invertebrates*, many of them microscopic.

When Floodplains miss out on Floods

When floodplains are left without floods for an extended period of time, nutrients, natural litter and salt accumulate. When a flood eventually occurs, a large amount of accumulated debris and vegetative material is washed from floodplains into streams as the flood recedes. This causes an overload of organic matter and de-oxygenation of the water due to mass decomposition. This is called a 'black water' event, as the de-oxygenated water usually looks black. With reduced frequency of minor flooding, black water events are likely to be more extreme when a flooding event finally occurs over long-dried floodplains. Environmental watering could be used to return more frequent small floods to floodplains to reduce the incidence of blackwater events.

Species which can be found in floodplain habitats: swamp lily, Peron's tree frogs and pigface daisy

Floods and Droughts in the Murray and Darling Rivers

Australia is a land of extremes and contrasts, known for varying between flood and drought. The Murray-Darling Basin is no exception, with periodic floods changing the landscapes and refreshing the ecosystems.

In the flood years, before weirs and dams, water spread out far across the floodplains, flushing backwaters and billabongs before draining back into the rivers. The process of inundating floodplains triggered a powerhouse of breeding and germination cycles, and when the water returned to the rivers, it carried nutrients, microscopic plants and animals, including young fish, which moved downstream.

In drought conditions, river ecosystems survived by reducing productivity and conserving energy. Plants produced less seed and slowed or even halted growth, while animals retreated to drought refuges like waterholes, or burrowed into damp soils. They survived on water stores that persisted from wetter times.

When the Rivers flood

In the Lower Murray below Mildura, most floods over the last 8000 years naturally occurred in October and November. This coincided with the highest chance of regional rain in August to November. River flows in the Murray system normally increased through spring as winter rainfall and snow-melt fed the rivers.

The largest recorded flood since European settlement occurred in 1956, with a peak flow in August 1956 of 350,000 Megalitres per day (*normal daily flow is 3-7,000 ML/day*). It inundated vast expanses of floodplains and river towns. First the Murray River and then the Darling River flooded, sustaining flood flows for nine months. 1956 was the second of two very wet years, so the storages were all full and the catchments were saturated, contributing to the wide extent and long duration of the flood. There was so much water that, even if the big upstream dams had all been empty, the flows would have filled all of them in four weeks, and still kept flooding for another eight months!

The narrow Murray Gorge flooded during 1993 floods

Key Points

- floods are a critical factor in sustaining life in river ecosystems, as water spreading over floodplains triggers breeding and regeneration cycles
- as floods recede, fish, aquatic animals and nutrients return to the rivers, continuing life cycles
- floods are inconvenient for humans but essential for nature to build soil moisture stores and resilience in ecosystems in order to survive the next drought
- droughts are also part of the natural cycle, likely to occur about once every 20 years
- the effects of dams, weirs and water extraction since European settlement have been to reduce floods and increase droughts
- rainfall has been decreasing at the same time, partly due to massive vegetation clearance across farming areas

Flood marker in the Lower Murray Valley shows the enormous height reached by the 1956 flood (top level), with many other historic flood levels below, but trees around the marker are stressed and dying from lack of big floods since the 1970s

Flood and Drought Sequences

Flows in the Murray-Darling Basin are naturally highly variable, often with several years of low flows, followed by several years of high flows. Big floods were recorded in 1870, 1917, 1931, 1945-46, 1955-56, 1973-75 and 2022-23. Small to medium flows occurred in 1989-1993 and 2010-12.

The years of 1955-56 recorded the highest floods since European settlement, but Aboriginal history reports even bigger floods in 1870 (when settlements along rivers were less extensive) and 1780. Interestingly, local Aboriginal stories describe the 1956 flood as the 'picanninnie' (baby) flood, with the larger 1870 flood and even higher 1780 flood as the 'mother' and 'father' floods. The two latter floods occurred prior to water records being kept and before extensive settlement of the river valleys, so there are few official records of their extent and impacts. Archaeological records include a red gum log lodged high in a cliff cave at Fromms Landing in the Lower Murray Valley by the 1780 flood.

The 1956 flood peaked at 50 times normal flows, and lasted for 9 months. As the 1955 flood was also a high flood year, all the catchments were saturated and dams were full before the 1956 flood peaked and river systems overflowed.

In 2022-23, three consecutive La Ninã events generated the second highest flood since European settlement, with uncontrolled spills when all the upstream dams were at full capacity. River levels rose over seven metres above normal in the Lower Murray Valley, where the flows from the vast catchment were squeezed into the narrow Murray Gorge in South Australia.

Floods the size of the 1956 flood have a 1 in 180 chance of occurring in any given year. Another flood of this size could still occur any time if there is another sequence of wet years. The chance of flood flows is independent of the previous years, so there can be a big flood two years in a row. Very often several very wet years occur consecutively, and this could still occur in future, with climate change increasing the chance of extreme events.

Key Points

- the highest chance of floods in the southern Murray-Darling Basin is in late spring to early summer, as rainfall and snow melt contribute to flows
- the highest chance of floods in the northern Murray-Darling Basin is in summer, when tropical monsoons bring rain to the catchment
- the largest recorded flood since European settlement was in 1955-56
- even larger floods are reported from 1870 and 1780, but with few records
- serious droughts occurred in 1895-1902 (the Federation drought), 1938-1645 and 2002-2010 (the Millenium Drought)

Significant floods in the mid-1970s in the southern Basin covered most inner floodplains and triggered regeneration and breeding but did not reach higher outer floodplains. In fact, some damage was caused by the weight of floodwaters pushing saline groundwater up to the surface beyond the edge of flooded areas, causing the death of river red gums which had germinated 20 years earlier in the 1950s floods.

A series of small to medium floods from 1989-1994 were beneficial in providing over-bank flows which inundated floodplains and triggered significant breeding and regeneration in animal and plant species, but these flows also did not reach the outer floodplains. The highest peak in late 1993 in particular was of great environmental benefit while causing only limited inconvenience to towns and settlements. A small flood in 1996 produced useful over-bank flows, but this was followed by 16 long years with no overbank flows reaching any floodplains. This river drought coincided with severe regional drought and lack of local rainfall, which left floodplains parched.

The next overbank flows did not occur until 2010-2012, when a series of flood peaks replenished soil moisture stores on the floodplain and refreshed water tables. These floods lasted long enough to trigger life-cycles, with mass breeding and regeneration events for all key floodplain species. A medium flood in 2016 provided useful overbank flows which watered floodplains, but its effectiveness was limited when high flows stopped suddenly in mid-summer and water poured off the floodplain, so there was no significant breeding or regeneration.

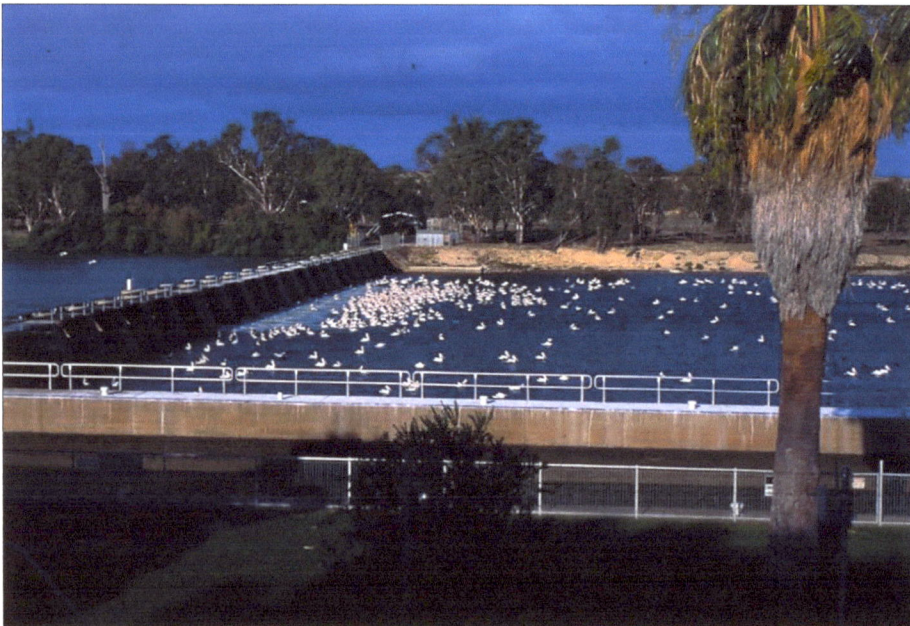

The weir at Blanchetown holds water levels up by three metres, one of a series of weirs which ensure a permanent flow. At the peak of the 2022-23 flood, the entire structure was underwater, with just the roof showing on the picnic shelter next to the lock chamber and the palm tree standing in water

In the floods of 2022-23, floodplains were inundated for many weeks, triggering significant breeding and regeneration events, as well as replenishing soil moisture and water tables. However, the speed and volume of flood flows drowned large numbers of the red gum and black box saplings which had germinated after the 2011 flood peak so it was a mixed blessing. The flood peak in the Lower Murray Valley was higher than the 1931 flood level but still not nearly as high as the 1956 flood level.

In between the flood years, serious droughts occurred in 1895-1902 (the Federation Drought), 1911-15, 1943-46, 1967, late 1970s,1982-83 and the Millenium Drought 2000-2010. The legacy of the extremely dry conditions in the Millenium Drought sadly includes millions of dead river red gums and black box trees on Murray Valley floodplains. The natural frequency of droughts is about 1 in 20 years, but the impact of water storage and extraction has effectively increased that to 1 in 2 years for the floodplains of the Murray River Valley. During low flows, there can be extended periods, even years, when no water passes the barrages at all and no water flows out through the Murray Mouth. Climate change predictions indicate that low flows will continue and there will be even less chance of natural flows to the sea.

Plants and animals of the floodplains and rivers are adapted to survive drought by mass germination and breeding in the wet times, and by relying on enough water being stored in soils and waterholes during wet times to be able to provide refuges in drought times. With water extraction and storage leading to fewer shorter floods, there has been less water available in droughts for plants and animals to survive. Millions of old mature trees, established in natural conditions long before European settlement,

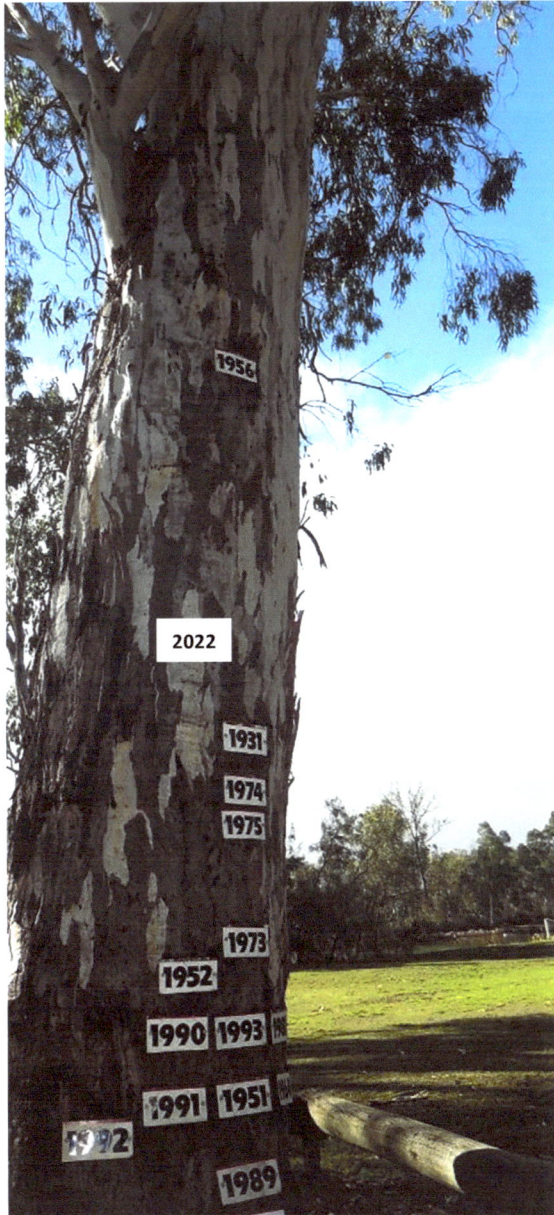

Flood levels on the 'Tree of Knowledge' at Loxton, showing the extreme 1956 flood level at the top marker, with the river in the background about 3 m lower than ground level at the tree base

Severe drought conditions in the Millenium Drought left millions of eucalypts along 700 km of the Murray Valley dead, dying or stressed. These trees were hundreds of years old, germinated long before settlement and had survived many previous droughts, but they had been water-stressed for decades before the Millenium Drought by high rates of water extraction

lived through all previous droughts but not the Millenium Drought. There simply was not enough water stored in their sap or in floodplain soils to support them through extended, very severe drought conditions.

Will the Rivers Flood again?

Since the weirs were built in the 1920s and 1930s, and large upstream storages were added in the 1930s to 1950s, the chance of flooding in October and November has been significantly reduced. At the same time, river operations have altered the timing of flow peaks, with higher river flows in summer and lower flows in spring. The chance of flooding has been pushed back to later in the year, when there is less chance of local rainfall and more chance of hot, dry conditions.

Under current managed conditions, floods are much smaller and shorter, because of the upstream dams and the amount of water pumped out of rivers. The flood in the summer of 1993-94 peaked at 126,000 Megalitres per day, but this was about half of the natural flows that would have occurred without dams. It also dropped away very sharply, instead of a gradual recession, reducing the success of breeding cycles. The 2010-12 floods peaked at nearly 94,000 ML/d (nine times regulated summer flows) and extended over 18 months, keeping floodplains wet and supporting mass germination of red gum and black box seedlings. A shorter flood in 2016 also peaked in late December at about 94,000 ML/d, but water levels fell extremely rapidly back to normal flows, again disrupting breeding cycles. The flood in 2022-23 peaked later, in early January, in the hotter summer months. It triggered breeding and regeneration of floodplain plants and animals, but they had to survive through hot, dry conditions after the water receded and survival rates to maturity may be reduced.

Annual flows depend on rainfall in the catchments. Every year, the chance of a high flow in late spring to early summer is assessed urgently, to determine how much water is allowed to be taken for irrigation and water supply, and how much water will be allocated for environmental flows. Contingency plans are made for flow scenarios ranging from flood to drought, ready to react and adjust as the yearly flow volumes develop.

The flood peak in 1993 at Kingston-on-Murray spread across the whole floodplain without serious threats to towns and major roads

Holiday houses along the river bank at Blanchetown were mostly elevated structures but it was not enough to escape a rise of seven metres in water level in the 2022-23 flood. All structures at ground level were completely inundated.

Will the Rivers ever Run Dry?

It seemed unthinkable that the Murray River could run dry. A minimum flow to South Australia was guaranteed by legal agreement from 1914 and the system of dams and weirs ensured permanent flows. The upstream managers of the river system are required to deliver 1,850,000 Megalitres (ML)[2] to South Australia every year, before they share the remaining water. Flows into South Australia range from 3000 ML/d in winter to 7000 ML/d in summer, with higher flows to deliver water for irrigated crops in the hot weather. Any flows above the required amount give an added bonus of extra flows in the river. In drought, the sharing rules change, and South Australia only receives one-third of available water.

During the Millenium Drought of 2000-2010, both the Mount Lofty Ranges catchment and the Murray River catchments failed at the same time, threatening Adelaide's water supply. The autumn rains failed in the Southern Basin catchment for nine years in a row. As severe drought conditions continued into 2007 and 2008, contingency plans were put in place to provide bottled water to Adelaide if Murray flows continued to fail, and a desalination plant was built on the coast.

Water levels in the Lower Murray dropped so low by 2008-09 that major sand banks appeared across the river channel. As river levels fell, river banks collapsed. Weirs and barrages had been holding the river permanently at a higher-than-natural level and when the level dropped, the saturated banks slumped into the river, taking trees and even cars with them. The Lower Lakes fell to 1 m below sea level, allowing salt water to push past the barrages, increasing salinity of the lakes and threatening Adelaide's water supplies. Emergency pipelines were built to supply Lakes towns and irrigated wine districts. Dairying was shut down on river flats and around Lake Albert. No water flowed to the sea so the Murray Mouth was dredged from 2002 to keep it open.

The town of Murray Bridge showing the extent of flooding in August 1956 (above, photo credit: News Ltd / Newspix), compared with normal regulated river flows in 1984 (below)

---

[2] 1 Megalitre = 1 million litres

Just before bottled water was needed for Adelaide, the extreme drought was broken by widespread rainfall in the upstream catchment in late 2010. River levels returned to 'normal', quickly followed by small to medium floods in 2010-12.

In recent years, it is the Darling River which has run dry. In 2015 and again in 2018-19, the Lower Darling River has run dry because of excessive water extraction for irrigation in the Upper Darling. This is of great concern for the towns and communities of the Lower Darling, as well as being critical for native fish populations of the wider Murray-Darling Basin which spend some part of their life cycles in the Lower Darling River. The plight of the Lower Darling came to world attention with dramatic pictures of mass fish kills in 2018-19, including large numbers of dead mature Murray Cod, up to 50 years old, the largest native fish in Basin rivers.

Communities throughout the Basin are calling for more sustainable management of flows in the Darling River to prevent future over-extraction from this important sub-catchment. Downstream users are very worried – if the Lower Darling can run dry, what guarantee is there that the Lower Murray won't run dry again, like it did in 2007-2009 during the Millenium Drought? The desalination plant in Adelaide remains to ensure water supply for the city during future droughts, but this would not provide water for river ecosystems and wetlands.

Rainfall is Declining and Run-off is Decreasing

Will the rivers run dry in the future? Will the river ever run freely to the sea again? The unknown factors for future floods and droughts are the combined effects of climate change and mass vegetation clearance which are reducing rates of rainfall and run-off into rivers. Rainfall over the Basin has declined significantly since the 1970s and run-off is predicted to reduce by 30-50%, giving much less water in rivers. At the same time, climate change brings an increased chance of extreme weather events, including short and intense flooding which may have the wrong timing for successful breeding and regeneration. The likely outcome is much less water available for users and for the health of rivers. River communities will need to plan for a future where there will be much less water available for extraction. Enough water must be kept in the rivers to maintain them in sufficient health for them in turn to support river-dependent human communities.

Normal river flows in 2006 with a full channel (above) contrasted with extreme low flows in 2009, which exposed sandbars halfway across the river channel (below)

Turning Variable Rivers into Permanent Rivers

After experiencing low river flows which stranded paddle-steamers and disrupted water supplies, early settlers set out to control the highly variable flows in the rivers of the Murray-Darling Basin. They wanted to ensure a permanent supply of water and year-round navigation by paddle-steamers. It took 30 years to reach agreement between the states and the River Murray Waters Agreement was signed in 1914.

A system of weirs, dams and barrages was designed by engineers from the US Army Corps of Engineers, who surveyed the rivers in 1905-06 to create a permanent navigable channel and to control the whole Murray-Darling system. Their designs included a network of weirs and fixed groynes at the Murray Mouth. Luckily, no ill-advised attempt was made to try to fix the high energy Mouth, but many of the proposed weirs were built, creating full regulation in the Lower Murray with 11 weirs from Mildura to Blanchetown. Some individual weirs were built further upstream at key off-take points to supply irrigation districts, for example Torrumbarry Weir to supply Shepparton and Yarrawonga Weir to supply the Mulwala Canal for the Deniliquin irrigation district. Construction on the first weir was commenced at Blanchetown in 1922 and the final construction was the barrages at the Murray Mouth, completed in 1940. By the time the weirs were completed the paddle-steamer trade had ceased, but the regulated flows and levels provided permanent water for towns and irrigators.

The weirs maintain water levels and store water for irrigation, rural and town use. The large upstream dams provide the water to maintain a summer flow in the river and to supply irrigators throughout the Basin. This system is essential to maintain permanent water supplies to Murray-Darling Basin communities. The barrages near the Murray Mouth exclude seawater to ensure adequate fresh water is available to supplement supply to Adelaide and regional South Australia, particularly during periods of drought, since South Australia lacks sufficient dam storage sites.

Hume Dam near Albury is the major storage on the Upper Murray, capturing spring flows for later release to supply downstream irrigators and towns

Key Points

- irregular flows in the rivers left paddle-steamers stranded, sometimes for 2-3 years
- after 30 years of discussions, a system of weirs, dams and barrages was designed to create a permanent channel for navigation
- 11 weirs were built from 1922, from Blanchetown to Mildura, to maintain this channel
- by the time the weirs were completed the paddle-steamer trade had ceased, but the regulated flows and levels provided permanent water for towns and irrigators
- 5 barrages were completed in 1940 to stop seawater from entering the Lower Lakes
- South Australia relies on Murray River water for domestic supplies during droughts, as there are not enough dam storage sites

On the Lower Murray River, 11 weirs from Mildura to Blanchetown control the river level in a series of stepped pools. This ensures that the river maintains a minimum level in summer. The weirs hold river levels three metres higher than the natural level and they are placed so that the effect is continuous. Any excess water flows over the top of the weirs.

Weir levels can be changed by removing or replacing *stop logs*, originally wooden logs and now concrete bars, which slide into slots on the weir pillars. During floods, all of the stop logs can be removed, leaving just the pillars with gaps between so the flooded river flows freely. The logs are replaced with very precise timing to ensure that the minimum water level is maintained after the flood recedes.

Weirs are usually referred to by the number of the lock chamber which allows boat passage through the weir (Weir & Lock No 4 is called 'Lock 4'). It is the weir structure that controls water levels in the river, not the lock.

The weirs are a barrier to the movement of animals through the rivers. This was belatedly recognised and a major program was finally completed in 2018 to allow native fish to migrate upstream past these barriers to spawn, using fishways specially designed to suit Australian fish.

Murray Mouth Barrages

As river flows reduced from the 1860s onwards, seawater intruded further into the Lower Lakes, affecting water salinity to the point where wool-washing at Narrung stopped in the 1870s because the water became too salty and drinking water supplies for lakeside towns and stock water sources were affected. A system of barrages was completed in 1940, using a limestone ridge formed by an earlier coastline as the base for the structures. The barrages were designed to keep salt water out, since the volume of water flowing down the River to the Mouth was now no longer enough to counteract tidal inflows.

Aerial view of Weir & Lock No 4 on the Murray River near Berri, with the lock chamber for boat passage at the right hand end of the weir

Weir & Lock No 2 downstream of Waikerie, showing the open lock chamber to the left, next to the navigable passage section which can be removed in floods to allow boat passage, and a mobile crane adjusting the height of the weir by removing stop logs between the concrete pillars

Lower Lakes

Since the construction of the barrages, the formerly mostly fresh Lower Lakes are now always fresh, supplying towns and districts on their shores, as well as ensuring freshwater supplies for Adelaide and many regions of South Australia.

The natural estuary was reduced to just 11% of its original area, restricted to the zone downstream of the barrages and only able to create a natural mix of fresh and salt water when river flows passed the barrages. In drought years, the Lower Lakes now act as a reservoir for several pipelines, ensuring that sufficient water is available to meet demands around South Australia. In low river flows, water can actually flow backwards from the lakes towards the primary pumping stations which lie between Mannum and Tailem Bend.

These wide shallow lakes are not ideal reservoirs, because a lot of water evaporates. Strong winds across the wide expanses can cause water levels to change dramatically, as the water is pushed to one side of the lake. However, water evaporating from the lakes contributes to rainfall cycles, so the water evaporating from their surface should not be classified as 'lost' but recycled.

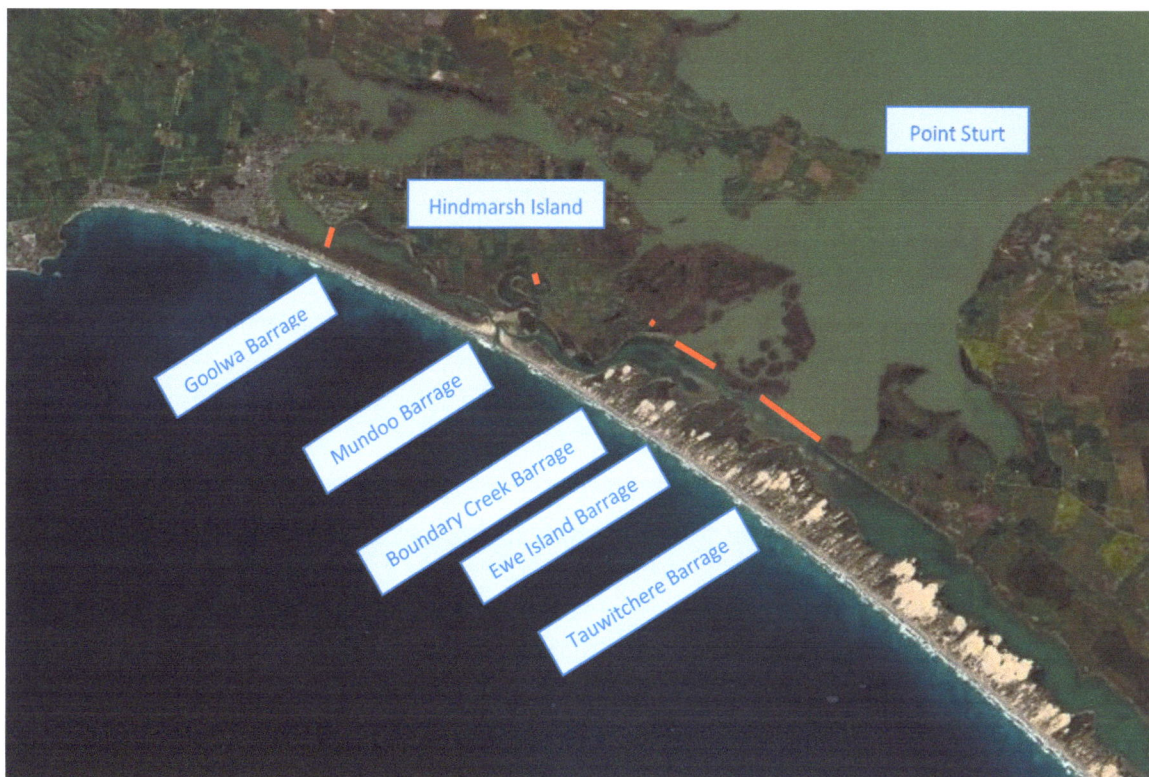

Aerial view showing the position of the five barrages which stop seawater from intruding into the Lower Lakes (base map: Nature Maps SA)

An Invaluable Resource

Water flowing to South Australia supports 16 river towns and five towns around the Lower Lakes, as well as more than 40,000 ha of irrigated horticulture. River-based tourism and recreation are major additional sources of regional income. This water is an essential back-up to domestic supply in dry times for 75% of the state population, including Metropolitan Adelaide. During low flows and regulated flows, there is no water left over to flow out to sea -- that is why the Murray Mouth has been dredged for 17 out of the last 21 years. Dredging keeps it from being closed by tidal flows pushing sand in through the Mouth. Sadly, the Murray River seems destined to join the fate of other over-allocated rivers world-wide which rarely flow to the sea.

The Murray River Boom and Bust Ecosystem

How does the Murray River ecosystem work? Over most of its length the Murray River is a slow flowing river with a very low gradient, carving its way across the inland plains west of the Great Dividing Range, meandering 2,300 km to the sea in South Australia. Even though it flows more swiftly in its upper reaches among the Snowy Mountains, the Murray River quickly becomes a convoluted meandering stream, winding through fertile valleys between mountain ranges and hills, then spreading across the flat plains.

On its way, the river meanders within a broad valley crossing semi-arid landscapes. Viewed from the air, the grey clays of the floodplains contrast with the red sand of the mallee country bordering the valley. River red gums line the banks of the mainstream and creeks and create dense clusters, particularly on the inner bends where there is sandy soil and more frequent flooding.

The natural, unregulated Murray River alternated between low flows contained within its channel in drier times and vast spreading floods, fed by rainfall in the Great Dividing Ranges and snowfall in the Snowy Mountains. The floods watered the floodplains and triggered mass breeding and regeneration of the plants and animals of the floodplain ecosystems. Groundwater tables were replenished in floods and helped to support plants and animals through dry times and droughts.

River flows support life in and along the Murray River. The way that water flows is the key to which plants and animals can grow in it or alongside it. This is called the *water regime*. It includes the range of fast flows, slow flows, pools, dry times and floods which spread out over river banks onto the floodplain. Flows can change in seasonal timing and from year to year. The ecology of the modern Murray River system has evolved over the last 8000 years to respond to smaller floods every 2-3 years, with larger floods every 5-10 years, and droughts only once in about 20 years.

*Wetlands on river floodplains respond to wetting and drying cycles to produce
microscopic plants and animals which are the basis of the food chain*

Key Points

- rivers flowing through semi-arid landscapes are hotspots of biodiversity and life
- rivers are interconnected with groundwater systems, replenishing them in wet times and drawing from them in dry times
- wetlands on river floodplains are biodiversity hotspots, where breeding and regeneration events are triggered when the wetlands are filled by high river flows
- Australian river ecosystems are adapted to 'boom and bust' conditions, breeding and regenerating in favourable flood conditions, and adopting survival strategies in drought
- native plants and animals are adapted to take advantage of wet times and to survive dry times, with a 'boom & bust' approach
- natural ecosystems of the Murray River have evolved in response to water regimes over the past 8000 years

Plants grow where they can get enough water to survive but not so much water that they would drown. Animals do the same, selecting places where water helps them to breed and survive, with food and shelter, and places to lay eggs and nurture their young. Fish find places in the river where the speed of water flow suits them, such as slower flows behind snags and deeper pools on the outer side of river bends. Increases in river flow speed and height will trigger fish to start spawning. Some species migrate upstream to breed, and the small fry drift back downstream to where their parents started out from. Golden perch (*also known as yellow belly*) have tracked swimming 2700 km from the Lower Murray River far upstream into the Darling River to spawn where they themselves hatched.

After 16 years dry, Nardoo emerged in the 2010-12 floods from the cracked dry floodplain, then persisted in scattered pockets through dry times in 2013-15 (left) and flourished again in the 2016 flood (right)

Wetting and Drying Processes

Australia is a land of extremes and its ecosystems are adapted to coping with the whole range of conditions from droughts to floods, typified by 'boom and bust' cycles of reproduction. Mass germination and breeding occurs in times of plenty, while plants and animals have various strategies to survive dry times.

The Murray River catchment is fed by autumn-winter-spring rains and snowfall, generating higher flows in spring. In the northern Basin, the Darling River and its tributaries are fed by summer monsoons. The wetlands along the Murray River Valley are geared to survive both spring floods and hot, dry conditions as water dries up in late summer and autumn. Plant and animal species have multiple methods for reproducing, usually including a low level of reproduction in most years, with mass breeding and regeneration in times of plenty of water. In flood events, water needs to flow over the mainstream banks and stay on the floodplain for 4-10 weeks to complete breeding cycles.

Importantly, wetlands need to be dry periodically, and then re-wetted, for the ecosystem to function properly. So Australian wetlands should not be wet all the time. Many wetlands actually need to dry out periodically for their ecosystems to work. In the dry phase, the dry, cracked soil stores eggs, spores, tubers, seeds, roots and other fragments of life in its deep cracks, ready to

In dry times wetland beds dry out into deep cracks which shelter seeds, eggs and animals; when the water returns, a new cycle of life starts, with microscopic plants and animals appearing first

burst into life when the water returns. Floodplain species of mussels, snails, yabbies and frogs survive by resting in damp soil until the wetland fills again.

When the water returns, a new cycle of life starts. Seeds germinate, dormant plants revive and flower, and eggs of fish, frogs and crustaceans hatch. The abundance of food attracts waterbirds, with ducks, swans and geese eating fresh green plants, and pelicans, cormorants and herons feeding on young fish and invertebrates. The life cycles are mostly completed in a few weeks or months, before the water dries up again, leaving eggs and seeds stored in the drying soil for the next cycle.

Floodwaters spread across floodplains, filling wetlands, creeks and shallow basins to create habitat for dragonflies, small fish and plants

Key Plants of Murray-Darling Basin River Systems

The Murray-Darling floodplains include wetlands, swamps, billabongs, creeks, anabranches, sandy river terraces and clay swales, red sandhills and claypan depressions. They offer a wide selection of habitats for many kinds of plants. Where different species grow indicates how often water is present in each location. Together, the trees and larger bushes provide protection for smaller shrubs and groundcover plants to create habitat and food sources on the floodplains for many animals, including frogs, birds, insects and reptiles.

The dominant plant species of the floodplains of the southern Basin are two eucalypts, the long-lived, deep-rooted river red gums and black box trees, and lignum shrubs. In the northern Basin, different eucalypts, grey box and coolabah, are dominant.

River red gums are found near permanent water or where floods occur every 2-3 years and semi-permanent fresh water and sandier soils occur. River coobah (an acacia which is sometimes called native willow) is also found close to the riverbank. Dryland tea-tree grows on the riverside sandy levee, and the rare prickly bottlebrush likes this habitat too.

Further back on the floodplain, shallow depressions are usually covered with lignum and sometimes old man saltbush. Open patches of grey clay support hardy tangled lignum bushes, which provide shelter for birds and other animals in dry times. In times of flood, lignum provides nesting sites for waterbirds and protection from predators for young fish. More salt and drought-tolerant than river red gums, black box trees grow at higher levels on the floodplain, in areas of clay soils which are less frequently flooded.

However, changes in water availability particularly since the 1950s are greatly reducing the numbers of seeds released and germinated, and the rate of survival to adult trees and shrubs. It is essential to take care of river red gums, black box and lignum, the guardians of our rivers, to ensure their survival, and the survival of habitat and food sources for so many dependent animal species.

River red gums line the rivers and creeks of Murray River floodplains

Key Points

- the dominant long-lived floodplain plants of the Murray Valley are river red gums, black box trees and lignum shrubs
- they are well-adapted to survive both floods and droughts
- river red gums require flooding once every 2-3 years for health and seedling production
- black box trees require flooding once every 5-10 years, plus local rainfall
- lignum shrubs require flooding every 3-4 years
- all three species provide habitat for many other species, including for insects, reptiles and birds
- black box trees are drought- and salt-tolerant, surviving on drier areas of the floodplain, living on lenses of fresher water over-lying very saline groundwater

River Red Gum Story

River red gums define the landscape character of the rivers and creeks of Murray-Darling rivers. Tall dense river red gum forests occur in the mid-Murray valley, where braided floodplains provide ideal growth habitat in the Barmah-Millewa and Pericoota-Gunbower Forests. Further downstream, more open river red gum woodlands grow on the Chowilla floodplain. Small pockets of taller red gum forest do occur, but only where frequent supply of freshwater is assured.

River red gums can be centuries old. Folklore suggests that each metre of circumference roughly equates to 100 years in age. The largest trees in the Lower Murray Valley are generally in the range 5-600 years old, with a grove of very old red gums near Morgan estimated to be 8-900 years old. These magnificent trees, often 30-40 metres high, grow from a golden seed smaller than a pinhead. Each tree sheds millions of seeds in its lifetime, but just one seedling needs to survive to replace its parent and to ensure survival of its community. The name 'red gum' comes from the distinctive internal wood, which is a rich red-brown colour.

River red gums need access to fresh water at least once every 2-3 years. Surprisingly, they do not take water directly from nearby rivers or streams, but access water in the soil or deeper groundwater layers via their root system of both shallow and deep roots. River red gums are quite tolerant of salinity; they can survive periods drawing on groundwater which is half as salty as seawater, provided that they get fresher water every 2-3 years. After five years without freshwater, the condition of red gums declines rapidly.

River red gums can tolerate standing in flood water for up to two years. They develop aerial roots above the water line to supply oxygen to the roots while the tree is flooded. However, after standing in water for more than two years, river red gums rapidly decline and die, as seen upstream of weir structures on rivers, where red gums have been standing in permanent water for decades.

Mature river red gums on the Packard Bend sandbar at Loxton in the South Australian Riverland

River red gums have mostly white bark with a dark grey base – they are named for the rich red-brown colour of their internal wood

Key Points

- river red gum trees can live to 5-900 years and are called 'nature's boarding house' because they provide food and habitat for so many species
- black box can also live for hundreds of years, with rough, dark bark retaining moisture in their trunks
- lignum is a long-lived, deep-rooted shrub which grows on clays soils on the floodplain
- red gums, black box and lignum hold their seeds on the plant, in aerial seed banks
- they release most seed when there is the greatest chance of seed falling onto moist soil, to germinate and survive through summer

Between floods, river red gums need periods when the soil is not flooded or water-logged, to allow seeds to germinate and develop, and for roots to access oxygen and to transpire. In drought, they conserve energy by reducing leaf, flower and seed production.

Nature's Boarding House

River red gums provide habitat for many species, earning the nickname 'Nature's boarding house'. Whistling kites nest up high, while kookaburras and butcher birds use lookout posts to spot their prey. Goannas can climb the trunks using their sharp claws to grip the bark. Their large hollows provide very valuable nesting sites for larger birds, including the vulnerable regent parrot.

River red gums provide nesting hollows, roosting perches and food for many birds and animals, including the pied butcherbird, darter chicks and goannas

On a smaller scale, geckoes and spiders hide under loose flaps of bark, and termites burrow underneath. Insects nibble on the edges of leaves, or skeletoniser insects remove the flesh leaving a skeleton of veins. Lerp (a sap-sucking insect) leave a series of uniform oval holes in leaves and hide their larvae under a white sugary cap. Small forest birds eat these caps, and so control the numbers of lerp.

Fallen red gum wood persists underwater for decades, providing hollow nests for native fish like Murray cod. Red gum leaves fall into the water year-round, providing food for aquatic insects specially adapted to graze on the tough eucalypt leaves.

Black Box Story

The second key eucalypt on Murray Valley floodplains is black box. They are found at higher elevations where they were naturally flooded once every 5-10 years. The oldest black box trees on the Lower Murray floodplain are thought to range in age anywhere from 200 to 1000 years. Their age is very difficult to estimate, as their growth stops and starts depending on water being available.

Black box trees do not tolerate standing in floodwater for long periods, so they form narrow belts of trees along the high-water line of floods. They can be divided into 'inner' and 'outer' communities, with the inner black box communities at lower elevations having a higher chance of flooding. Outer communities on the higher floodplain where floods are less frequent rely on local rainfall between floods.

Black box trees are typical of the 'box' group of eucalypts, with dark grey-black, rough bark on the trunk and branches. They are generally half the height of river red gums, about 10-15 m tall. The taller trees grow where they have better access to water and trees are much smaller on the higher, drier floodplain. Black box trees which germinated in the 1956 flood and have not been flooded since are still very slender and only 3-4 m tall.

Black box trees are very tolerant of salinity; able to survive on groundwater which is as salty as seawater, provided that they get fresher water from either rainfall or floods every 4-5 years. Like red gums, their root system is a combination of shallow and deep roots, to access water in the soil or deeper groundwater. They rely on rainfall between floods to recharge unsaturated soils and freshen the groundwater. However, reduced flooding has left large areas of higher floodplains without floods to freshen saline groundwater for over 40 years, resulting in devastating losses of mature black box trees which were hundreds of years old.

From Tiny Seeds to Majestic Trees

In the Murray Valley, both river red gums and black box mostly flower in summer, November-December, although some red gums and black box flower as late as March. The timing is thought to be linked to the likelihood of floods in late spring-early summer. A minority of black box trees flower in winter, May-July, instead of summer, apparently relying on winter rainfall to provide soil moisture. The timing of flowering sets the timing for seed fall 12 months later. River red gums and black box trees hold their seeds in aerial seed banks on the tree for up to two years, before dropping most seed when there is a high chance of sufficient soil moisture to germinate seedlings and support them through the following summer.

Dead black box where no floods have occurred for over 40 years (above); healthy mature black box trees and seedlings at a location where groundwater is fresh (below)

Golden river red gum seeds are the size of a pin head (above), and are hard to separate from chaff, as seen under an electron microscope (below)

The millions of seeds produced by river red gums provide food for other species, with ants the most common hunter. Only relatively few seeds germinate, although thousands of seedlings may appear in one location when conditions are right. A river red gum is estimated to produce 600,000 seeds each year. One healthy river red gum in the Riverland was measured producing 995,000 seeds. The chances of successful germination are estimated at 5%, or potentially 49,750 seedlings from this one tree, so nature has ensured a good chance for survival of this species. Black box trees produce about one-seventh of the seeds compared to river red gums but their seeds are much larger and easy to distinguish.

The key for survival is for seeds to fall on bare moist soil, with enough moisture to support the seedling through the following summer. If seeds fall onto water, they can be swept to the edge by wind and then germinate along the high-water mark, creating lines of red gum saplings around wetlands and along river banks, or black box at the outer edge of floods. Seedlings need to be protected from stock until their tips are above grazing height.

Red gums develop new leaves every summer, while shedding large numbers of older leaves. They also shed a lot of bark, covering the ground with a thick crunchy carpet which provides shelter for insects, reptiles and other small animals. Black box also renew their leaves every summer but do not shed their rough bark.

Black box seeds are dark and striped, easily separated from golden chaff under a microscope

Tangled Lignum Story

The humble tangled lignum, usually overlooked in the shadow of river red gums and black box trees, is the workhorse of the floodplains. Lignum is a long-lived, deep-rooted plant providing valuable habitat. It is a survivor with high tolerance to both drought and salinity. In dry times, its thickets provide shelter to many small animals and plants, such as fairy wrens and seedling red gums. Flooded lignum provides a nesting platform for waterbirds like ibis and freckled duck. It also shelters young Murray cod, golden perch and other young native fish from water currents and from predators.

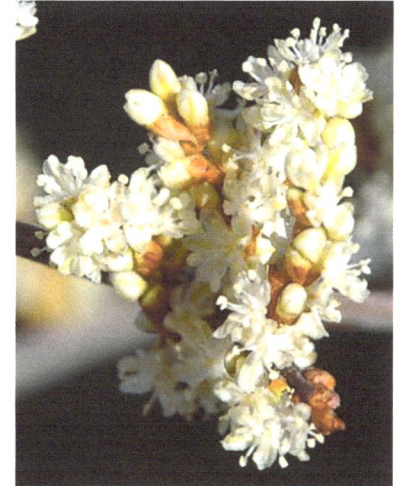

Flowers of river red gums (left), black box (centre) and male lignum (right)

33

Lignum plants grow to 1-3 m height, with a rootstock at least 2-3 m deep. The tangle of growth starts with vigorous vertical stems extending 3-4 m, until they arch over under their own weight. New growth starts from each node on the stems, sending a new set of vertical stems up, before they inevitably arch over, compounding the tangle.

During dry times, lignum goes dormant, looking like a tangle of dead brown sticks. It starts a new life cycle within a few weeks of rainfall events or floods, with greening of stems, then new leaves and flowers. Leaves appear first but only last for a few weeks; they are shed after flowering. The flowering season is short, 3-4 weeks, and

Microscopic photos of lignum flowers, female (left) and male (right) showing the green star- shaped female flower cluster and the exerted stamens of the male flower

seeds form 4-6 weeks later. Ripe seeds drop from the plant and need to fall on moist soil or water to germinate.

Lignum has separate male and female plants. The male flowers, with eight prominent fertile stamens, are more obvious. The yellow- green female flowers are smaller, with five petals and a tri-branched style, held close to the stems in tight clusters. Lignum can reproduce both from seeds and from vegetative growth. Both methods require abundant water for success, either from rainfall or floods. In wet times, lignum can stand in water for several months. Arching stems can eventually land on damp soil and roots growing from stem nodes can create new plants. Shoots can also grow from nodes on roots underground, creating straight lines of green vertical shoots around the parent plant.

Lignum is an exceptionally useful plant which fulfils multiple roles. In addition to its habitat value, its deep root system helps to keep saline groundwater below the root zones of river red gums and black box. It covers large areas of river flats, shallow clay pans and nutrient-poor cracking clays on the floodplains, reducing evaporation from the surface.

Lignum flowering in response to environmental watering (left); healthy lignum in flower after spring rains (right)

Animals of Murray-Darling Rivers

Many different kinds of animals are associated with Murray-Darling rivers, including birds, mammals, fish, turtles, lizards, snakes, frogs and a host of *invertebrates* (animals without back bones). The mammals and birds may be more obvious and appealing, but all species are part of the same life-support system which keeps the river healthy, catering for all the plants and animals, even the smallest invertebrates.

The river's edge—a biological 'hot spot'

The river's edge and wetlands are home to most of the river animals, including mussels, snails, shrimps and yabbies. Even large fish like Murray cod rely on this rich zone for food, as a refuge, or as a nursery. By comparison, relatively few kinds of animals are found in open river channels.

Plants like bulrush (or cumbungi), reeds, sedges and ribbon weed grow in a narrow fringe along the river's edge and around the edges of wetlands (the *riparian zone*). These plants provide food, shelter and breeding sites for the animals of the rivers and their floodplains.

A special feature is the presence of snags (large pieces of fallen timber), either on the river bank or in the water. These provide shelter and surfaces that allow animals and plants to attach, including hollow logs underwater for fish nests, sheltered pockets of still water out of currents and surfaces for biofilms.

A family group of young emus on the Chowilla floodplain

Many animals use the floodplains and rivers

Kangaroos and emus graze widely across the floodplain. Both swim well, so don't be surprised to meet them mid-river. Possums are found in the trees and a wide range of woodland birds, including many parrot species, can be observed.

Snakes and lizards are common, and also frequently swim across the river. As soon as water reaches any low plants on the floodplain, frogs will be heard calling. Sightings of the threatened southern bell frog (or growling grass frog) have increased in the southern Murray-Darling Basin during recent flood events.

Brown snake swimming across the main channel of the Murray River

Key Points

- rivers and floodplains provide habitats for many different animals
- the river's edge and wetlands are biological 'hot spots', key locations for breeding and feeding
- snags are important in-water habitat
- fallen timber, twigs and bark are important habitat on land
- mussel species are adapted either to swift-flowing water in the river channel or still waters in floodplain wetlands
- crayfish like swift-flowing waters while yabbies like still waters
- 17 out of 18 native snails have disappeared from the Lower Murray Valley

A Tale of Two Mussels

Mussels are among the most abundant large invertebrate animals in the rivers and their shells are common along river banks and near wetlands on the floodplain. Live mussels burrow in the sediments down to about three metres' depth and live in shoals of 2-3 per square metre.

The larger river mussel grows to about 18 cm and survives for 30 years or more. It lives only in the big rivers of the Murray-Darling system and needs water that is cool, well supplied with oxygen and above all, flowing strongly. The billabong mussel, which grows to about 12 cm and has a lifespan of about 15 years, is found in billabongs, lakes, swamps and minor streams throughout eastern Australia. It tolerates high temperatures and is able to survive when oxygen falls to very low levels. It can live out of water for a year or more – a key adaptation for an animal living in the periodically dry wetlands of the Murray River floodplains.

Mussels Depend upon Fish

The life-cycle of mussels depends on native fish to host their larvae (*glochidia*). These attach to the fins or gills of the host fish for three weeks. Young mussels then emerge and drop to the river bottom for a few years before joining the adult mussels burrowed along the river's edge or in deeper water. Each female mussel produces many thousands of tiny glochidia in spring or summer, but the odds for survival of an individual glochidium are very small. They must attach to a suitable host fish and then survive predation by bottom-feeding fish like catfish and carp.

The river mussel (above) has a distinctive extended shape, designed to help anchor it in swiftly flowing water, compared to the rounded billabong mussel which lives in still or slow-flowing water (below)

Goannas can swim across the river and make use of emergent snags while hunting prey

Shell Middens

The mounds of bleached mussel shells along the banks of the river are middens, or refuse heaps, left by the original Aboriginal inhabitants of the Murray Valley. The middens may be thousands of years old, and are a priceless record of the way the river was. The Aboriginal groups would gather mussels from the river and billabongs and roast them in their cooking fires. Over the generations very large mounds of shells accumulated. The changing proportions of shells in the middens suggest that billabong mussels have become more common since European settlement, most likely due to the effects of water extraction and weirs.

Disappearing Molluscs

Fossil oyster shells in the cliffs that flank the narrow Murray gorge are part of the geological layer known as the North-west Bend Formation and a legacy of times when the sea flooded the valley, perhaps 6 million years ago. Oysters are molluscs, like their relatives the mussels and snails. The fossils of many other marine and estuarine invertebrates can also be found in the river cliffs.

Dense fossils of oysters and other marine life found in cliffs

The molluscs of modern times are freshwater animals and the Lower Murray Valley was home to 18 species of snails. However, since settlement and river regulation, 17 species are now rare or extinct. Snails feed on biofilms, the slimy growths of algae, bacteria and fungi found on rocks, wood and plants submerged in the water, but the nutritional value of biofilms has reduced in the more stable water levels of regulated rivers. Only the tiny freshwater limpet remains abundant in the river, thought to be able to survive because it feeds on tiny particles in the biofilms that are not accessible to other snails. The tiny conical limpet shells (the size of a pin head) are found on the broad, flat leaves of ribbon weed in shallow water along the river's edge. The decline of snail species is a clear indicator of ecosystem decline and a stark warning of declining river health.

Another factor in the decline of snails is the effect of abrupt daily changes in water level due to weir operations, and the extremely rapid drop in water levels of 1-2 m overnight as floods recede and weirs are reinstated. Most snails are unable to adjust and are left stranded. The disappearance of an entire group of animals, in this case the gastropod molluscs, is a sign of major change in the ecosystem. Further disruptions are caused by competition from an introduced European snail and the destructive effects of the introduced European carp on snail habitat. The decline of snail species is a clear indicator of ecosystem decline and a warning that all is not well in the Lower Murray Valley.

The Murray yabby breeds in large numbers when floods spread out over floodplains

Ironically, one river snail which had disappeared from river environments was found to have become a pest in regional irrigation pipelines, breeding up in high numbers due to nutritionally richer biofilms inside the pipes. These river snail populations are being relocated into local wetlands to try to re-establish wild populations.

A bucketful of yabbies caught during a flood

Crustaceans of the Murray

In the modern Murray Valley there are two kinds of crayfish: one is the familiar smooth-shelled yabby, the other is the spiny Murray crayfish. The yabby, like the billabong mussel, is a hardy species that tolerates high temperatures and low oxygen, and is well adapted to life in billabongs, swamps, creeks where the water is not flowing swiftly. The Murray crayfish is like the river mussel: it needs deep, cool, well- oxygenated water, and lives in the main channel of the river and its bigger tributaries.

Other crustaceans are prawns and shrimps. Dominant male common prawns, up to 10 cm in length, have long, pincer-bearing front legs. Smaller shrimps (1-2 cm) are the most abundant of larger invertebrates in the rivers. Together, these two species are important as grazers on biofilms, and as a key source of food for fish. Prawns are also opportunistic and will feed on dead animals.

Water Rats

Native water rats or *rakali* are rodents, known by their white tail tips. Small piles of broken mussel shells lying on a flat rock or log out in the river are a sign that water rats are around. The shells have small holes made by their sharp incisor teeth. Water rats are common but very shy and most active at dawn, dusk and at night. Mussels are one of their staple foods, and they also eat yabbies and fish.

Turtles of the Murray River

The smaller eastern long-necked turtle (above) and larger broad-shelled turtle (below)

Three species of turtle, all aquatic and all freshwater, inhabit the Lower Murray River. These are the small eastern long-necked or snake-necked turtle and the two much larger species, the broad-shelled turtle and the short-necked turtle. Turtles emerge from the water to nest in banks along the river and wetlands. The eggs are sought by many predators, including foxes, water rats, goannas and ravens, and plundered turtle nests are a common sight.

Warning sign at Tortoise Crossing floodway, a traditional migration path

The smaller eastern long-necked turtle is typically an inhabitant of swamps, oxbows, billabongs, and slow-moving rivers. Extensive overland migrations sometimes occur during summer. The eastern long-neck feeds on molluscs, crustaceans, tadpoles and small fish, and lays up to 10 or more eggs in a hole dug on the bank of its home stream or swamp, usually in early summer. The broad-shelled turtle is essentially a river turtle and is harder to see in the turbid water. This species lays eggs in autumn and the young hatch out in spring. The short-necked turtle is restricted to larger river channels and larger waterholes on the floodplain. A summer breeder, it lays around 10 eggs in a chamber dug high on the river bank. It feeds on crustaceans, molluscs and fish.

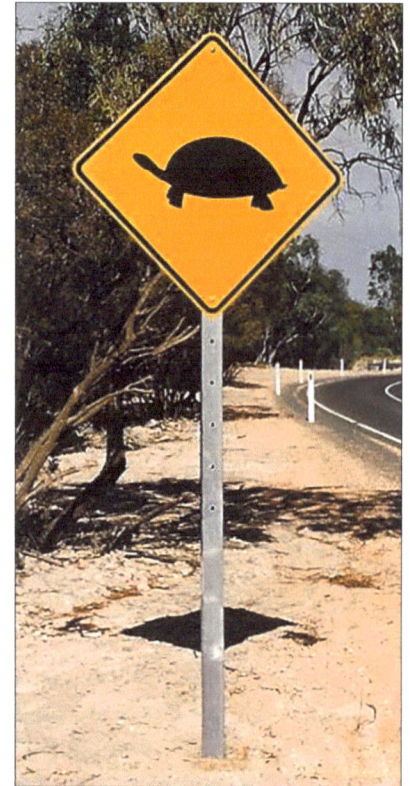

Frogs of the Murray River

The presence of abundant frog populations indicates healthy wetland and river systems. Eight species of frog occur in wetland and floodplain habitats along the Lower Murray River. They include tree frogs, swamp frogs and froglets.

Peron's tree frog

Peron's tree frog (or maniacal cackle frog) occurs along the length of the Murray River and may be heard calling between September and January. The brown tree frog occurs in the lower reaches of the river in South Australia, and may be heard calling all year. The southern bell frog (or growling grass frog) is the largest tree frog and calls from August to April. This frog was thought to be in danger of extinction in the Lower Murray Valley, but recent observations indicate that it is locally common in the Riverland.

The eastern banjo frog or pobblebonk occurs along the length of the Murray River and is heard calling all year round. Two similar looking frogs are the barking marsh frog and the spotted grass frog. Both occur along the length of the river in South Australia, but while the spotted grass frog calls year-round, the barking marsh frog generally calls between October and March.

The eastern sign bearing froglet is found along the length of the Lower Murray Valley, while the common froglet is restricted to the lower reaches. These froglets, no larger than 3 cm long, call year-round.

The main threats to frog survival and recruitment are insecticides, loss of habitat due to draining or alteration of wetlands, and the introduction of gambusia (misnamed *mosquito fish*) which prey on frog eggs and tadpoles but are not effective at controlling mosquito larvae.

Shallow flooded plants in late spring make perfect frog habitat

Birds of the Murray-Darling Basin

The rivers and wetlands of the Murray-Darling Basin are home to more than 130 species of Australian native birds, both forest birds and waterbirds[33], which use the habitats of waterways, floodplains and woodlands.

Straw-necked and white ibis roosting

The Murray River winds its way from the alpine meadows of the Snowy Mountains, down through cool green hills to the dusty inland plains, then continues its meandering way to the wide expanses of the Lower Lakes, the bleached sandhills of the Coorong and the roaring seas of the Southern Ocean. Along the way it feeds thousands of creeks, lagoons and billabongs, providing food and shelter for waterbirds of all shapes and sizes. The Darling River flows through the arid interior, providing a green ribbon of life through semi-arid shrublands and mallee woodlands, intermittently filling large lake systems in wet years to support key fish nurseries and waterbird rookeries.

The rich birdlife has evolved to ebb and flow with the droughts and floods of the river system, breeding in large numbers during the boom times of floods and finding different ways to survive droughts. In dry times they can move to wetter regions or retreat to drought refuges.

Waterbirds rely on floodplains and wetlands for breeding and survival. These provide suitable nesting, roosting, feeding and loafing sites. Nesting sites vary from the cliff ledges to hollows in living and dead trees. Breeding is triggered in response to flooding and related food sources. If flooding persists, some species nest two or even three times in one season.

Waterbirds found across the Basin include the soaring pelican, as well as herons, huge flocks of ducks, and large waders like avocets and stilts. Some of the important nesting colonies which occur along Murray-Darling rivers include sacred and straw-necked ibis, royal and yellow-billed spoonbills, great white egrets and several species of cormorants. Other species are less conspicuous, but still very dependent on healthy river ecosystems. The nankeen night heron, with its distinctive hunched form and glorious cinnamon (nankeen) colour, is a seasonal visitor. It migrates between northern

Key Points

- the Murray-Darling Basin is home to more than 130 Australian forest and waterbird species
- birds can adapt to varying conditions, flying to flooded breeding grounds and retreating to drought refuges
- important feeding grounds around the Lower Lakes and Coorong are visited every summer by migratory waders which fly from as far away as Siberia
- waterbird populations have been steadily declining since 1983
- populations have not recovered when intermittent floods have occurred, indicating the devastating effect of drought periods and reduced flooding

a well-disguised tawny frogmouth

---

[3] the term **waterbirds** describes any bird that displays a strong association with water in its life-cycle

Australia and southern wetlands in response to rainfall. They may be seen roosting in trees overhanging the rivers by day and foraging for food at dusk. The Coorong hosts many species of ducks, swans & geese, as well as tiny migratory waders which fly every summer from Japan, Korea and Siberia to the Lower Lakes and Coorong.

Regent parrot feeding its young

Smaller forest birds using the floodplains include rainbow bee-eaters, honey-eaters and wrens. The gorgeous rainbow bee-eater can often be seen perching high above the rivers and swooping to catch insects at dusk. They have blue and green plumage, with translucent ochre on the underside of the wings and tail shafts extending the length of the tail. They nest in a burrow in sandy ground, often on a riverbank, cliff or cutting.

Rainbow bee-eater watching for insects

The threatened freckled duck is a master of disguise, with its mottled brown colouring camouflaging it well amongst its preferred habitat of lignum swamps. It breeds mostly in irregularly flooded lignum swamps in the north-west Murray-Darling Basin, and then flies into southern regions as the floods dry up. It listed as threatened because of declining population numbers.

Forest birds are also in decline. The nationally vulnerable Regent parrots, with their brilliant yellow plumage, streamlined design and black wing tips, are exciting to watch as they dart through the red gums with shrill calls. Numbers in the Lower Murray Valley have reduced to a small population in the Riverland, thought to be about 400 birds. Prior to being protected, they were shot by fruit-growers as pests because they were attracted to stone-fruit crops. The major problem for them now is the lack of large hollows for nesting sites, with thousands of mature red gums stressed and dying. A special project funded by concerned Adelaide citizens purchased water during the Millenium Drought to irrigate stressed mature red gums at a key breeding ground for Regent parrots at Hogwash Bend near Cadell in South Australia[4].

Superb fairy wren (Photo Tom Hunt)

---

[4] Environmental watering project of **Water for Nature** program of Nature Foundation SA in 2008-2009

Movement Patterns in Birds

Migratory birds found along the Lower Murray include the rainbow bee-eater, fairy martin and Caspian tern. Caspian terns banded from a large breeding colony near Mount Isa in Queensland have migrated south as far as Murray Bridge in South Australia, a direct flight of about 1500 km. Some of the nomadic birds found on the Lower Murray include the black-tailed native hen, little corella, white-faced heron, black kites and grey teal. Australian reed warblers are commonly heard along reedy waterways of the Lower Murray from September through to March, before they migrate to warmer regions each winter. Australian grebes are found in shallow lagoons and creeks next to the mainstream in the Lower Murray during the cooler months between May and September.

Migratory shorebirds which travel from as far away as Siberia are mostly found around the Lower Lakes and Coorong.

Cliffs of the Lower Murray Valley are important nesting sites for many birds

Birds of the Lower Murray Cliffs

Great white egret

Towering limestone cliffs are a feature of the Lower Murray River in South Australia, extending from Overland Corner to Mannum. They are a haven for birdlife. They offer excellent protection both from the elements and predators, and are used by many birds as nesting and roosting platforms. Ledges on the cliffs are important breeding sites for peregrine falcons and nankeen kestrels. Large colonies of Sulphur-crested cockatoos are also found in the cliffs.

Other birds that use the cliffs for roosting and nesting include the sacred kingfisher, laughing kookaburra and barn owls. Fairy martins and welcome swallows build mud nests attached to the undersides of cliff ledges.

Food Sources for Birds

The diet of birds within the Murray-Darling Basin varies greatly according to species and available food. Most carnivorous species are willing to feed on a variety of animals. Depending upon availability, they will take crustaceans, insects, other invertebrates, amphibians, small reptiles, mice, birds and eggs. Other bird species are predominantly herbivores: they feed on aquatic biofilms, water lilies, sedges and submerged plants like water ribbon. Floodplain plants are another source of food, with flowers and seeds, as well as hosting other food sources like insects.

The period following high water levels offers ideal breeding conditions for insects and invertebrates. Waterbirds such as herons and ibis take advantage of this episodic food source. Receding floods stimulate the reproduction of many fish species, providing food for fish-eaters such as pelicans, darters, grebes and cormorants. Spoonbills feed on fish, insects and crustaceans.

Black swans at a freshwater soak near the Murray Mouth

Seasonal conditions can result in a mouse or grasshopper plague and many birds will utilise these as their major food source. Barn owls and kestrels appear to thrive during a mouse plague. The nankeen kestrel feeds mostly on insects such as grasshoppers and beetles, but the bulk of its food its food is small mammals. Cockatoos and corellas feed on seeds, including spilled grain.

Waterfowl of the Murray-Darling Basin

Waterfowl is the term for ducks, swans and geese. They are mostly plant-eaters (*herbivores*) and have webbed feet. There are 19 native waterfowl species in Australia, 16 of which occur in the Murray-Darling Basin.

Most waterfowl breeding in the Murray-Darling Basin follows the flooding of wetlands, but some breed at the same time each year regardless of flooding. Australian waterfowl mostly breed when dry wetlands re-flood, because inundation stimulates the production of their food (aquatic grasses, water lilies, sedges and submerged plants including pondweeds, nardoo and duckweed). Some species also feed on aquatic insects and invertebrates.

Waterfowl found in the Murray-Darling Basin include:

- Hardhead or white-eyed duck
- Freckled duck (the least abundant Australian duck)
- Grey teal – the most numerous Australian duck
- Chestnut teal
- Blue-billed duck (the male has the blue bill)

Pink-eared ducks (Photo Tom Hunt)

- Black swan (the largest Australian waterfowl)
- Pacific black duck (often sedentary, very common through the Murray-Darling Basin)
- Australian shoveler (common in the Murray-Darling Basin)
- Pink-eared duck (a nomadic bird, most common in the Murray-Darling Basin)
- Maned duck or wood duck (a grass feeder that has benefited from agricultural activities)
- Magpie goose (mostly limited to northern Australia; uncommon in the Murray-Darling Basin)
- Australian shelduck or mountain duck (common in the Murray-Darling Basin)
- Plumed whistling duck (uncommon in the Murray-Darling Basin)
- Wandering whistling duck.

Australian pelicans roosting

Australian pelicans

A flock of Australian pelicans swimming slowly in formation is a peaceful and fascinating sight. Pelicans swim in coordinated groups to drive fish into shallow water so that they can scoop them up into their bills. They dip their bills into the water simultaneously, withdrawing them slightly open to allow water to drain out of their pouches. If they have been successful, they toss up their heads and swallow the fish.

They are nomadic birds and will frequent any suitable body of water. Pelicans breed in colonies, building a nest of sticks and plants on a rough scrape on the ground. Pelicans soar on thermal air currents, an inspiring sight. For many, their annual breeding ground is in the Coorong. However, opportunistic mass breeding will occur when inland salt lakes fill with freshwater during floods. When Lake Eyre fills, there can be up to three consecutive breeding events, although the third event is likely to fail as the lake dries and food supplies for the chicks decline before they are old enough to leave the nest.

Australian pelicans are found throughout Murray-Darling rivers and wetlands

Cormorants and Darters

Cormorants and darters are diving birds that spend much of their time beneath the water searching for food. Their plumage does not repel water, giving these birds a drenched or soaked appearance. This in fact is an advantage to a diving bird, as the resultant lack of buoyancy decreases the effort required to dive. After diving, cormorants and darters are frequently seeing drying out their wings in the sun, to enable them to fly again.

The Future for Birds in the Murray-Darling Basin

Waterbird numbers across the Basin are in steep decline due to loss of habitat. Many wetlands have been lost to drainage or changed by river regulation, so that less food is available and fewer chicks survive. Relatively common species such as black swans, black duck, grey teal, chestnut teal, white ibis, straw-necked ibis, wood duck, yellow-billed spoonbill and small waders have all declined significantly since 1983. Without environmental flows which mimic natural flows and flood patterns to support breeding and survival of Basin waterbird populations, this decline will continue.

Australasian darter (Photo Tom Hunt)

Royal and yellow-billed spoonbills feeding (Photo Tom Hunt)

Red-necked avocets feeding in the Coorong

Native Fish of the Murray-Darling Rivers

Compared with big rivers in other parts of the world, Murray-Darling rivers have a very low number of native fish, only about 30 species. In fact, there are few native species of fish in all of Australia because of the long geological isolation of the continent and the frequently arid climate. A few very ancient species like the Queensland lungfish have survived. However, most freshwater fish are, in evolutionary terms, recent invaders from the sea.

Young Murray cod (photographed and returned)

The Murray cod is the best-known native fish in the Murray- Darling system. Other familiar species are catfish, silver perch, callop and bony herring, but only the last two are still common. Smaller native fish still found include galaxias (native trout), western carp gudgeon, big-headed gudgeon and congolli. Small fish like blackfish and pigmy perch are now very rare.

Curiously, there are no eels in the Murray, although they are found in most other coastal rivers of southern Australia. This may be because the oceanic currents which carry young eels from their breeding grounds do not pass the Murray mouth. Eels are not to be confused with lampreys (jawless fish), which look like eels but have a sucker mouth to attach to other fish. There was great excitement when a lamprey was caught in 2017 as far upstream as Renmark!

Common names are confusing. The Murray cod is not a true cod and the bony bream is a herring. The Aboriginal names would be far more suitable for uniquely Australian fish – the cod is *ponde*, the golden perch is *callop* and the catfish is *tandan*.

Key Points

- only 30 species of native fish occur in in Murray-Darling rivers
- the largest native fish is Murray cod, other large fish include golden perch, catfish and silver perch
- small native fish including purple- spotted gudgeon and carp gudgeon are an important part of the food chain for larger fish and waterbirds
- native fish are struggling to survive and breed with changed flow conditions and barriers to migration
- native fish are struggling to compete against invasive species like European carp and gambusia
- invasive European carp now form 80% of fish biomass in Murray-Darling rivers
- introduced gambusia (misnamed mosquito fish) are not as good as native fish in controlling mosquitoes

Catfish, photographed and returned

Life Cycles of Native Fish

A flood benefits all river fish. It must be at the right time of the year and the water level must remain high long enough for fish to move onto the floodplain to spawn, for their eggs to hatch and baby fish to grow. Even fish that do not move into flooded areas benefit from the invertebrates that grow in flooded wetlands and are carried back to the river as food for young fish.

Murray-Darling fish have very special breeding mechanisms based around the highly variable natural flow patterns. Some fish spawn only when there is a flood, while others spawn every year hoping for a flood to provide sufficient food for fry to survive. Some fish species migrate upstream to breed and the juveniles travel back downstream. Their breeding depends on floodwaters persisting on the floodplain for long enough to reach warm enough temperatures to trigger proliferation of microscopic plants and animals, creating food sources.

The life cycles of fish involve floods in different ways, a bit like race betting:

- *Sure thing* – callop and silver perch spawn only when there is a flood
- *Two bob each way* – rainbow fish and carp gudgeons spawn seasonally, whether there is a flood or not, but more intensely when there is a flood
- *Double or nothing* – Murray cod and catfish also spawn seasonally, independent of floods, but floods provide much more food for the young and increase survival rates
- *No bet* – lampreys, galaxias and congollis breed in estuaries or the sea and young fish return to the river.

Shelter is important too. Floodplain bushes and fallen timber provide protection from currents and predators. Cod have been shown to shelter inside flooded bottlebrushes, while lignum is a key plant for providing pockets of still water and shelter for fish fry and fingerlings. Cod lay their eggs inside hollow logs underwater.

Purple-spotted gudgeon, one of the small native fish species of the Murray-Darling rivers (Photo K Walker)

Professional fisherman's catch on a Murray backwater creek in the 1980s included one large Murray cod, several callop and many carp (the ones with large scales). There are no longer sufficient native fish to support a professional fishery.

Changes in River Fish Fauna

The range and abundance of many Murray fish have changed in the past 200 years. Some, like the bony herring and carp gudgeon, may have increased. Others, including Murray cod, have done less well and some species like Yarra Pygmy Perch are approaching extinction. The principal changes that have affected native fish are alterations in flow patterns resulting from the effects of dams, weirs and water extraction, interactions with introduced species and perhaps past fishing practices.

Big floods can still occur nowadays but the effect of dams and weirs has meant that smaller floods are much less frequent than they once were. Whilst cod, callop and other fish breed extensively in big floods, they also rely upon smaller floods to breed on a small scale. In this way, the numbers of fish in the population are maintained so that, when a big flood does arrive, there are sufficient fish to take advantage of it. The reduction of smaller floods may be partly responsible for the long-term decline of some native fish. Murray cod spawning and survival is optimum if there is a big flood at least one year in seven, with smaller floods in between.

Lower Darling River is Critical Habitat for Murray-Darling Fish

Only recently it was found that the Lower Darling River is critically important fish habitat for the whole Murray-Darling Basin, as many large native fish spend some part of their life cycle in the Lower Darling and then migrate to other rivers. The Lower Darling and Menindee Lakes are the primary nursery areas for golden perch (callop) populations, with young fish migrating from there to other Basin rivers. The Lower Darling River is also one of the key nursery sites for Murray cod.

Unfortunately, flows to the Lower Darling River have been steadily reduced since 2013 by very high volumes of water extraction in the Upper Darling. Lower Darling flows have been so low that it was unable to withstand drought and algal blooms. The Lower Darling River ran dry in 2015-16 for 500 days, creating great distress for both human and fish communities. It dried again in summer 2018-19, with catastrophic effects. Three tragic fish kills decimated native fish stocks. Thousands of mature Murray cod up to 50 years old died, along with other large native fish like callop, as well as millions of smaller native fish which provide food for larger fish and waterbirds.

After the drying event in 2015, environmental flows were directed to the Lower Darling and stimulated the best cod breeding event for 20 years. However, most of those young cod were lost in the 2018-19 drying events and catastrophic fish kills. It is likely to take decades for native fish to recover from these mass deaths. Securing flows to the Lower Darling for the future is the solution. Waterholes and refuge areas need to be maintained during droughts and sufficient flows ensured to trigger and support breeding when the rains come. Environmental flows which aim to mimic natural river flow patterns and small floods give the best hope for the survival of native fish.

A black water event occurs when flooded vegetation or accumulated leaves and bark decompose, leaving no oxygen for aquatic animals and turning the water black

Effects of Black Water Events

A serious threat to fish health can occur with a 'black water' event. This happens when a large amount of accumulated debris and vegetative material is washed from floodplains into streams as a flood recedes. This causes de- oxygenation of the water due to mass decomposition of this overload of organic matter.

With reduced frequency of minor flooding, black water events are more extreme when a flooding event occurs over dry floodplains. A major black water event occurred in 2016 when Lachlan River floodplains were flooded after 20 years without any inundation. This water flowed from the Lachlan River all the way to the Lower Murray River, causing fish kills, particularly in backwaters where fish could not escape the anoxic water. Environmental water was used to provide a freshwater refuge flow for fish alongside the black water. Long term environmental watering can return more frequent small floods to reduce accumulated organic material on floodplains, resulting in less blackwater events.

Effects of Introduced Species on Native Fish

Ten fish species have been introduced to Murray-Darling rivers and are now naturalised there. They include brown trout, rainbow trout, redfin perch, gambusia, goldfish and carp. Other invaders include aquarium escapee oriental weather loach and tilapia. Each of these has been responsible for significant changes and have wreaked as much havoc underwater as cats, foxes, goats and rabbits have on dry land.

The introduced fish are better adapted to the changed, regulated Murray River, with its stable water levels and reduced floods. Brown trout and Redfin perch are predators. Gambusia (misnamed mosquitofish) was introduced in the 1930s and 1940s to the Pacific region by the US Army Corps of Engineers to control malaria. Unfortunately for the Murray-Darling Basin, which had no malaria, gambusia are not as good as the local native fish at controlling mosquito larvae, and they eat the eggs of native fish and nibble the fins of adult fish. Gambusia are only small, 2-3 cm long, but they breed quickly, attaining large numbers in a short time. They give birth to live young, and are very tolerant of environmental extremes, including high salinity. Native species like pigmy perch and rainbowfish are voracious mosquito predators, much more effective and preferable to the destructive introduced gambusia.

The worst damage is caused by European carp, which now make up 80% of the fish biomass in the Murray-Darling Basin. In the 1960s a wild strain of carp was imported illegally from Germany to a fish farm in Victoria. They were spread deliberately through farm dams in the Murray catchment, until massive floods in the mid-1970s swept them into the rivers. They spread rapidly and quickly reached very high densities, and are now harvested by commercial fishermen for crayfish bait and food markets.

Carp have caused serious damage to wetlands along the river valley. They feed by sifting silt on the bed of wetlands to find microscopic animals, uprooting water plants and muddying the water. Carp can be controlled in local areas, using fish screens to exclude adult carp and drying cycles to kill them before they reach breeding age. Control of water levels can reduce opportunities for spawning, or desiccate eggs after spawning. Wetting and drying wetlands can consolidate the wetland bed, making it harder for carp to feed.

Where carp were excluded by a screen from one half of Little Duck Lagoon, the water plants grew strongly (left) but where carp were allowed in, they stirred up the bed, displaced the plants and made the water muddy (right)

Recent proposals to control carp are based on introducing a lethal virus, reported to be specific only to carp, but this is now being questioned. In addition, careful management measures would be needed to manage mass simultaneous carp deaths and to remove huge quantities of dead fish which could rot and severely deplete oxygen in river water. The result could be vacuum in the ecosystem which would most likely be filled by other pest species, as not enough native species are likely to breed over a short period to re-populate fish habitats.

Why Wetlands are so Important

Wetlands in the semi-arid Murray-Darling Basin are priceless assets. They are essential resources for the driest inhabited continent and continued survival and well-being of the Basin is dependent on their health.

Everywhere on earth wetlands are being destroyed, yet they are essential to the ecological well-being of the planet. If they disappear, life for all living things is more difficult – animals, plants and humans. Wetlands are not competitive water users, they are essential– their well-being is our well-being.

Wetlands in Australia

At settlement, natural wetlands in Australia covered more than 240,000 km². After 250 years of water diversions, drainage and dams changing river flows and drying wetlands, less than half survive at all. There has been severe loss of diversity and health. In high value agricultural areas, as little as 1-2 % of the original area remains.

Australian wetlands are highly variable, never average! In the land of 'droughts and flooding rains', natural cycles dictate that wetlands are sometimes dry, sometimes wet. Native plants and animals which rely on these wetlands are adapted to the 'boom and bust' conditions, alternating between droughts and floods.

In order to survive the droughts, rivers and wetlands need to be allowed to flood in wet times. If the floods are cut off from floodplains and stored in dams or diverted to irrigation, the rivers and wetlands will have less capacity to store moisture reserves needed to survive the next drought.

Key Points

- wetlands are hotspots of biota in river ecosystems
- wetlands in Australia need to have wet and dry cycles
- wetland functions include filtering, trapping sediment, converting nutrients into food and flood buffering
- wetlands in Australia's highly variable climate are adapted to floods and droughts
- wetlands have boom breeding and regeneration events in wet times and can shut down to conserve energy in dry times
- wetlands need the right amount of water at the right time persisting long enough for life cycles of plants and animals to be completed.

Temporary wetland at Teschner Bend near Renmark, ringed by healthy, very old river red gums in 2000

Perons tree frogs are quick responders to wetlands filling, with a deafening chorus at dusk and into the evening

Wetlands in the Murray-Darling Basin

The wetlands of the Murray-Darling Basin are essential to the well-being of all human communities dependent on the waters of the Basin. Healthy Basin wetlands mean healthy communities and water-dependent industries sustained for the future.

Wetlands of the Murray-Darling Basin need to be flooded regularly but they also need to dry occasionally to be healthy. Drying wetlands out lets air into the soil, cracks the mud, firms the wetland bed and frees nutrients. Wetting a dried wetland starts up the food chain, turns nutrients into plants and animals, starts breeding cycles and new seedlings, starting a new cycle of life.

Floodplain plants and animals respond rapidly when water spreads onto the floodplain, creating habitat and food sources

Keeping wetlands healthy

Healthy wetlands are like Goldilocks: they need conditions to be 'just right', just enough water, often enough, for long enough, at the right time, and of the right quality. If they get these conditions often enough, they will support water plants, water birds, forest birds, migratory birds, native fish, crayfish and yabbies, tortoises and frogs, as well as healthy plant fringes around their edges. They will filter out sediment and nutrients, keeping river water clean and clear, and slow down floods. Murray-Darling wetlands need enough water to thrive so they can provide these essential ecosystem services to Basin communities!

In the regulated Murray River, particularly downstream of Mildura, 30-40% of wetlands are now permanently full, drowned by higher river levels behind a series of weirs which maintain pools to supply towns and irrigation settlements. These wetlands can benefit from managed drying phases, where it is possible to exclude water using water control structures, and then re-fill wetlands to mimic the natural timing of temporary floods.

Continuing threats to wetlands in the Murray-Darling Basin include:

* removing too much water from rivers
* altering timing, length and frequency of flows
* no small or medium floods reaching wetlands on the floodplains
* not enough water to maintain low flows and refuge waterholes in drought
* increased blue-green algal blooms and black water events
* introduced pests competing with native species and favoured by slower flows and stable water levels (carp, willows, foxes, aquatic weeds)
* loss of fringing vegetation from river and wetland edges.

Waterbirds like Royal spoonbills respond to flooded wetlands which provide a combination of nesting and roosting sites close to food sources

Why Water Plants are Good Guys

Water plants are the basis of the food chain in wetlands and rivers, providing food and shelter for other plants and animals. They feed many waterbirds, and provide material to build nests. They provide habitats for frogs to feed and breed, sheltering eggs and tadpoles. They host microscopic plants and animals which provide a food source for native fish and crustaceans. They provide shelter from predators and shade from the sun. The floating fern *Azolla* creates a floating carpet, starting out green and gradually turning red as it is exposed to the sun. Underneath, microscopic animals hide in its roots and tiny fish hide from predators.

Water plants also stablilise the bed of wetlands and river banks, reducing erosion and allowing sediment to settle out of the water column. They convert nutrients in the water column into plants and animals, leaving the water clear and less likely to develop algal blooms. So, water plants are 'good guys' and should never be described as 'weeds'. It is important that water plants are protected and retained in our waterways to keep our rivers and wetlands healthy.

Dragonflies responded in thousands when floodwaters spilled onto floodplains in October-December 2016

Important water plants include water ribbon half-submerged in pools (left), nardoo which emerges from flooded floodplain clays (centre) and azolla, a floating fern which provides shelter at the surface from sun and predators (right)

Where the Murray River Meets the Sea

The end of the long journey for the Murray River is the Lower Lakes, Murray Mouth and Coorong. The last little town on the river mainstream is Wellington, before the River discharges into two large shallow terminal lakes, Lake Alexandrina and Lake Albert. Beyond Point Sturt on the southern side of Lake Alexandrina, the main river channel winds around Hindmarsh Island. Other channels find their way through a chain of sand and mud islands into the Northern Lagoon of the Coorong. All the channels eventually reach the Murray Mouth and the Southern Ocean. The shape and location of the Murray Mouth has varied over geological time under the influence of the forces driven by the river, wind and sea. Historically, the Mouth was kept open by high river outflows that periodically scoured out sandbanks which had built up from beach sand pushed in by the tides.

Lakes Alexandrina and Albert, the Lower Lakes

Lakes Alexandrina and Albert were predominantly fresh water ecosystems for several thousand years pre-1850, with an estuarine boundary near Point Sturt on the southern-most edge of Lake Alexandrina.

Evidence from the traditions of indigenous Ngarrindjeri people, scientific analysis of sediments, historical records of plants and animals, with their known salinity tolerances, lake-shore shell middens and computer modelling of water exchange all show that the Lakes were fresh to very slightly saline for the last 7000 years. Cores from the centre of Lake Alexandrina showed no marine diatoms were present. Large persistent populations of freshwater mussels and freshwater fish species found in lakeshore middens indicate sustained freshwater conditions.

Seawater intruded only occasionally into the main body of Lake Alexandrina prior to European settlement, because flows from the Murray River usually were sufficient to counteract tidal inflows. While salt water could enter under extreme low Murray River inflows, these were short-lived episodes and never approached salinities of seawater.

Since construction of the barrages, the Lake ecosystems have adjusted over the past 80 years to permanent freshwater conditions and higher, more stable water levels. In recent years, fishways have been built and barrage gates operated to assist movement of fish at critical times in their life cycles.

Ewe Island barrage showing low seawater levels in the Coorong

Key Points

- the Murray River discharges into Lakes Alexandrina and Albert, known as the Lower Lakes
- the Lower Lakes were fresh when first settlers arrived around 1840 and scientific evidence confirms they were fresh to very slightly saline for the last 7000 years
- the natural estuary, where river water mixed with sea water, was downstream of Point Sturt
- by 1870, settlers were noticing increasing salinity in the Lakes due to upstream use of water
- eventually, five barrages were built in 1939-40 to exclude sea water and maintain the Lakes as fresh for towns and irrigators
- Murray River flows are now less than 40% of natural flows and the barrages are essential to keep sea water out of the Lakes
- the main river channel flows west around Hindmarsh Island to reach the Murray Mouth, while other channels link to the Coorong
- the Murray Mouth has been dredged to keep it open 2002-2010, 2013-2016 and again since 2017 due to river outflows being too small to push sand out

Natural Murray River Estuary

Estuaries are transitional zones between rivers and the ocean. The sea influences the estuary through tides, waves and inflows of salt water. The river influences the estuary through large inflows of fresh water, sediment, salt, nutrients and other materials. Estuaries are highly productive communities, with a diverse mix of plant and animal species including both transient and permanent residents. They are critical areas for fish nurseries and waterbird feeding and breeding.

The natural Murray estuary was located in the southern parts of Lake Alexandrina, with a mixing zone south of Point Sturt most of the time. The estuarine zone was located primarily in the channels all around Hindmarsh Island but the boundary with fresh water would move,

Aerial view showing the linkages between the Murray River, Lower Lakes, Murray Mouth and Coorong
(Base map: Nature Maps SA)

however, in accord with changes in river flows, wind and tides. The modern estuary has existed for about 7000 years, after sea level rose rapidly from about 120 m below sea level 18,000 years ago to its present level.

The local indigenous Ngarrindjeri regard the Murray estuary as 'The Meeting of the Waters', an area where *ngartji* (totems) breed and the waters of the river, tributaries, lakes, Coorong and sea are mixed and infused with spiritual energy. According to traditional knowledge, the boundary of the estuary was in the Goolwa Channel south of Point Sturt.

Tauwitchere barrage is the longest, with a central set of gates modified to provide cues for fish to migrate past the barrier.

The salinity gradients, remnant estuary and connection to the sea mean the wetland complex of the Lower Lakes, Coorong and Murray Mouth represents a unique wetland system in the Murray-Darling Basin. Waters varying from fresh to brackish to salty and super-salty, combined with tidal mud flats and sand banks, attract many waterbirds and migratory shorebirds which fly from as far away as Siberia to feed in the Coorong every summer.

Murray Mouth

There are five channels from Lake Alexandrina to the sea: Goolwa channel is the deepest and carries most of the flow, circling to the west of Hindmarsh Island. Mundoo Channel (or Holmes Creek), Boundary Creek, Ewe Island and Tauwitchere are shallow channels with limestone sills of old shoreline deposits (*these sills provided the base for the barrage structures built to exclude sea water*). The four latter channels discharge into the Northern Lagoon of the Coorong and then flow either towards the Murray Mouth or upstream into the Coorong, depending on tides and wind.

Dredge operating at Murray Mouth to clear sand build-up in 2006

The shape and location of the Murray Mouth vary in a dynamic interplay between river, wind and sea. The Mouth has migrated over 1.6 km since the 1830s and some 6 km in the past 3000 years. The Murray Mouth rarely closed in the past 3,500 years, including those between 1837 and 1981. However, with weirs, dams and increasing extraction of water, marine tides now dominate over river outflows and there is a much higher chance of closure. Sand deposits inside the Mouth have formed islands which have gradually blocked channels inside the Mouth.

The Mouth closed in 1981 for the first time since European settlement in 1837, and closed again in 2002. Dredges have operated ever since, except for pauses during flood flows in 2010-12 and again briefly in late 2016. Dredging has continued from 2017 into 2022 due to very low flow conditions, which are likely to persist under climate change predictions. Dredging has started again in late 2023, post floods.

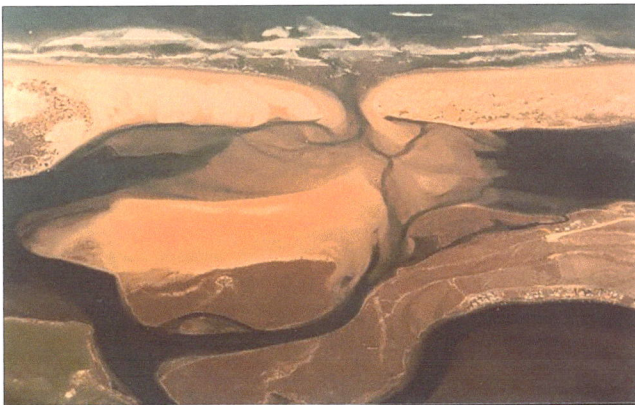

Murray Mouth barely open in 1983, assisted by dredging

Murray Mouth was well open in 1984 with strong river outflows from both Goolwa channel (right) and Tauwitchere channel (left); the Mundoo channel is closed (lower right)

Murray Mouth in 2015, with very restricted channels kept open by dredging

The Coorong, Wetland Jewel at the End of the Murray-Darling System

At the end of the Murray-Darling system, the path to the sea is not straightforward. Water splits into multiple channels through a barrier of sand and mud islands to reach the Southern Ocean. While the main river channel winds to the west of Hindmarsh Island, past Goolwa, other channels to the east reach the Coorong Northern Lagoon before flowing either downstream towards the Murray Mouth or upstream into the Coorong, depending on tides and wind.

The Coorong stretches for about 100 km upstream from the Murray Mouth. The Northern Lagoon, nearest the Mouth, is brackish to salt (*estuarine to marine*), depending on how much freshwater is flowing in from the river. The Southern Lagoon, separated from the Northern Lagoon by shallow, restricted zones at 'The Narrows' and 'Hells Gate' at Parnka Point, is *hypersaline*, from 1.5 to 4 times as salty as seawater.

The Coorong is an elongated coastal lagoon lying in the swale between two coastlines, the current coast and the previous coastline from about 16,000 years ago. It is unusual, in that it is a 'reverse estuary', with freshwater inputs from the Murray River at the downstream end, rather than from the upstream end as occurs in a normal estuary. Freshwater is pushed upstream through the Northern Lagoon by strong wind effects (*seiching*) which mix the water and push it through the restricted connection into the Southern Lagoon, which stretches further south-east past the little township of Salt Creek.

The Murray River is the primary source of fresh water flowing into the Coorong. Historically, flows from the South East only rarely reached the Coorong via Salt Creek at the southern end, and only in the second of two wet years, when enough water had accumulated across the South East region. The water flowed slowly downhill towards the Coorong, finding its way through a series of swamps and creeks to cut across parallel sand barriers of past coastlines. High points in flow paths were removed to increase flows from the South East and a 4 m deep drain was cut through the inland dune at Salt Creek in 1864. Even with assisted flows, Drainage Board records show

Typical sandhills of the Younghusband Peninsula, separating the Coorong from the sea, with a fringe of vegetation supported by freshwater soaks

Key Points

- The Coorong is a shallow coastal lagoon about 100 km long
- Its main source of freshwater is the Murray River
- This makes the Coorong a reverse estuary, with its freshwater source at the downstream end
- It takes high river flows and strong north-westerly winds to push fresher water into the Southern Lagoon
- The Northern Lagoon is about the same salinity as seawater or fresher, while the Southern Lagoon is saltier than seawater
- With the Lower Lakes and Murray Mouth, the Coorong is a wetland of international significance, providing varied habitats for tens of thousands of waterbirds
- Tiny migratory waders fly from Japan, Korea and Siberia every summer to fatten up on the rich mudflats
- Pelicans have their only annual breeding colony here

water only flowed into the Coorong six times over a 48-year period to 1912. Natural flows into the Coorong from the South East were only a rare secondary source of fresh water.

Further south in the Coorong swale, in the remnant of the geologically historic estuary, the ephemeral lakes are separated from the Southern Lagoon and rely on local rainfall and run-off combined with rising groundwater to fill them in winter. As a result of the annual wetting and drying process, many are actively precipitating carbonate minerals, including dolomite, magnesite and calcite. These lakes support important populations of salt-tolerant plants around their fringes and in the water, providing key food sources for waterbirds.

Wetland of International Significance

The Coorong is a magical place. It is very important in Aboriginal legends, and home to many very special creatures. The Coorong, Lakes Alexandrina and Albert, and the Murray Mouth are listed as a wetland of international importance under the Ramsar Convention. The value lies in the mixed habitats of freshwater lakes, fringing saltmarsh, marine Northern Lagoon, hypersaline Southern Lagoon and the mineralised ephemeral lagoons. It provides a haven for waterbirds and can be a drought refuge when the rest of their habitat is dry.

The Coorong is separated from the ocean by a continuous barrier of complex sand dunes, the Younghusband Peninsula. There are frequent freshwater soaks along the western edge of the lagoons. Rain falling on the sand dunes trickles out into the soaks, creating small oases of fresh water in a saline environment. Specialised plants on the dunes and in the shallow water provide a variety of foods to support large flocks of waterbirds.

The mixed habitats of the Coorong, Lower Lakes and Murray Mouth combine to provide habitat and food sources for a wide range of species, particularly thousands of migratory waders which travel from Japan, Korea and Siberia every summer to fatten up on the rich food sources

Northern Lagoon of the Coorong, looking upstream with Ewe Island and Tauwitchere barrages to the left of the picture.

Narrow passage between the Northern and Southern Lagoons known as 'Hell's Gate' at Parnka Point

A chain of ephemeral lakes extends beyond the Southern Lagoon of the Coorong, filling seasonally in winter and drying in summer to create unique mineral deposits

Red-necked stints fly from Siberia every summer (Photo: Tom Hunt)

These tiny birds eat busily every day to stock up sufficient energy to make the return journey to their breeding grounds, a trip of more than 10,000 km. Japanese snipe can be spotted visiting salt marshes bordering the western shores of Lake Alexandrina at Milang every summer.

The Coorong supports regular breeding of Australian pelican, fairy tern, crested tern, caspian tern and silver gull. Pied oyster catchers, red- capped plovers and chestnut teal breed occasionally. Important water plant communities including musk grass and sea tassel provide food for duck species. The Cape Barren goose is a regular summer visitor, with large flocks fattening up on grassy paddocks across Hindmarsh Island before returning to their breeding islands off the South Australian coast in winter.

The pelican is particularly associated with the Coorong, maintaining its population with a breeding colony at Jacks Point on the Southern Lagoon. Some pelicans have bred here every year for as long as can be remembered. This is the back-up insurance plan, to ensure some annual breeding in between the irregular rare breeding events when inland saline lakes like Lake Eyre fill with fresh water. The pelican illustrates the 'boom and bust' ecology seen so often in Australian ecosystems, taking advantage of inland floods for mass breeding events. Many other species have

Southern Lagoon near Salt Creek, looking across to the sandhill barrier

dual breeding strategies like the pelican, to make the most of times of plenty of water and food in the variable Australian climate.

The hypersaline Southern Lagoon attracts waders like avocets and stilts, feeding on myriad worms and crustaceans, as well as small salt- tolerant hardyhead fish. The shores are lined with deep drifts of tiny pink Coxiella snail shells.

Freshwater inflows from the Murray River enter the Northern Lagoon of the Coorong and mix slowly into the hyper-saline Southern Lagoon under the influence

Red-necked avocets (Photo: Tom Hunt)

Adult pelicans at the Coorong breeding colony (Photo: Fiona Paton)

of winds and tides. Freshwater needs to be added at the downstream end to maintain the natural salinity gradient. The lack of flows from the Murray River in recent years have led to rising salinities in the Coorong beyond natural tolerances of the unique ecosystems, particularly during the Millenium Drought. In 2006, the Southern Lagoon became so salty that there was a bloom of brine shrimp, and small numbers of banded stilts and avocets bred. However, salinities continued to rise and the brine shrimp and birds disappeared. The floods of 2010-12 and again in 2022-23 were critical in bringing salinities back to more tolerable levels.

Over recent years, increased flows have been directed into the Coorong from the South East as a result of drainage works to manage rising groundwater in that region. However, concerns are being raised about the impact of these small, frequent inflows into the upstream end of the Coorong, as mass blooms of filamentous algae have developed, smothering key food plants in the Southern Lagoon.

It will be critical for the future to ensure that sufficient fresh flows are delivered from the Murray River to sustain the internationally significant wetland complex, to maintain water quality and natural levels in the Coorong and to keep the Murray Mouth open to export salt and toxins out of the Basin.

Banded stilts feeding in the Coorong (Photos: Fiona Paton (left) & Tom Hunt (right)

Pair of fairy terns (left) and fairy tern chick (right) (Photos: Tom Hunt)

How have the Rivers Changed?

The rivers of the Murray-Darling system were naturally highly variable in their flows, reflecting the highly variable climate which produced everything from floods to droughts. Measures implemented to control flows and support navigation and irrigation started in the 1920s. Large upstream dams, mid-stream storages and a series of weirs have largely been managed to ensure permanent flows in the main channels and enough stored water to sustain supply to towns and irrigators. The river regulation system has captured so much water that flood events are much smaller and less frequent. At the same time, continued water extraction has left rivers, effectively, in man-made droughts. The lack of flooding has reduced the capacity of river ecosystems to survive drought by limiting their opportunities to replenish water stores.

The system of 11 weirs from Mildura to the Murray Mouth has turned the Murray River into a series of stepped lakes, with each weir raising water levels by 3 m. The water flows more slowly, so species which prefer swiftly-flowing water have declined, while floodplain species which prefer still water have migrated into the main channel. River mussels have been replaced by floodplain mussels, and Murray crayfish have been replaced by yabbies. The timing of flow peaks has changed to later in the season, with higher flows in summer to supply irrigators and lower flows in winter and spring, changing breeding cues for fish, frogs and waterbirds.

Wetlands no longer experience the natural cycles of wetting and drying which followed the natural pattern of floods. Where there are weirs holding river levels higher, wetlands close to the rivers are permanently wet, so they do not dry out occasionally as they once did. These wetlands are marked by large fields of dead trees, particularly red gums, drowned in permanent water. Wetlands higher on the floodplain do not fill as often as they once did, leaving them dry and salty, with many trees at these sites dead from lack of freshwater and too much salty groundwater in their root zones.

As much as 80% of the water is diverted away from the rivers, with serious impacts on the health of the floodplain, its wetlands and the river itself. Since it has been regulated, the Murray River no longer reduces to low flows in summer.

River levels are raised by 3 m at each weir, so water backfills into adjacent wetlands, keeping them permanently full

Key Points

- naturally highly variable flows have been controlled by dams and weirs
- rivers now run permanently, with much reduced flooding
- large volumes of water are extracted from rivers for irrigation and town supplies
- changes in the last 200 years are reducing suitable flood events which can support breeding and regeneration events for waterbirds, fish, frogs, turtles and plants
- flood peaks are later and shorter, occurring in hotter months, so there is less water to support growth and survival of seedlings and young animals
- natural systems are losing their resilience and finding it harder to survive droughts
- accumulating salt and nutrients need to be flushed out of the Basin through the Murray Mouth
- a minimum amount of water needs to be recovered for river health, so Basin rivers can continue to support Basin communities

It also does not flood as often, so birds, fish and plants do not breed and regenerate as regularly as they once did.

River ecosystems are adapted to survive droughts, but they do that by replenishing groundwater during floods and having mass breeding and germination when there is plenty of water to enable seedlings and young animals to survive to adulthood. With so much of the floodwater captured in dams, the river ecosystems struggle to survive droughts. In the Millenium Drought, millions of river red gums and black box trees died. These trees were hundreds of years old, but could not survive extended, severe drought conditions because they had been deprived of floodwaters in previous years which would have replenished moisture in their root zones.

After the floods of 2010-12, some very stressed red gums put on new growth, but millions were already dead from the severe Millenium Drought 2000-2010

The numbers of native plants and animals have been reduced and some struggle to survive, even to the point of local extinctions. River ecosystems have lost their resilience, their ability to survive through drought, because the wet times have been reduced and soil moisture reserves and water tables are not being replenished. Millions of trees have died and new seedlings struggle to survive and grow to maturity.

Photographs of the river with almost no water have been widely published, with people standing astride a tiny stream or a Sunday School picnic being held on the river bed, but these images date from dry times after irrigation settlements were already drawing large volumes of water out of the river to sustain orchards and vines. Fish biologists have found evidence that there were always low flows in Basin rivers, even through drought periods. Large long-lived native fish like Murray cod would not have been able to survive if rivers were reduced to a series of isolated water holes. There must have been minimum flows to keep them alive.

Native fish have been blocked from migrating upstream to spawn by weirs built across the rivers. Where spawning has occurred, young fish returning downstream are often killed trying to swim over these weirs. Fish-eating birds know fish are frequently stunned and pop up to the surface just below a weir, so flocks of cormorants and pelicans can be seen lining up to catch them. Over recent years, fish ladders suitable for Australian fish have been designed and introduced to the weirs, to allow the native fish to pass through. Sadly, it has taken nearly 100 years to understand the needs of the native fish and to modify the weirs so they can complete their spawning migrations.

The construction of weir pools along the Lower Murray in the 1920s and 1930s allowed yabbies to invade the river channel from the floodplain. They became very common in the river, until the early 1970s and the arrival of carp. They are still common at times, especially after a high river that brings food and opportunities to disperse and breed. The Murray crayfish is virtually extinct in the Lower Murray, but still occurs further upstream. Small numbers may remain in the Lower Murray in deep holes in the river channel near Morgan, and although numbers sometimes increase briefly after good flows, the crayfish has been declared an endangered species. If caught, they should be returned to the water!

The groundwater underlying the Murray River from Swan Hill and downstream is naturally as salty as the sea. This salty groundwater flowed slowly towards the river and the salt was carried out to sea. Salt loads in the river have increased as irrigation has mobilized salt in soils and transported it into rivers. Water table mounds under irrigated areas exert pressure on the groundwater, speeding up its flow under floodplains towards rivers. Nutrient levels in rivers are also higher due to run-off from agricultural uses. More and more water is being taken out of rivers, reducing flows and causing problems of water quality, particularly more frequent toxic algal blooms.

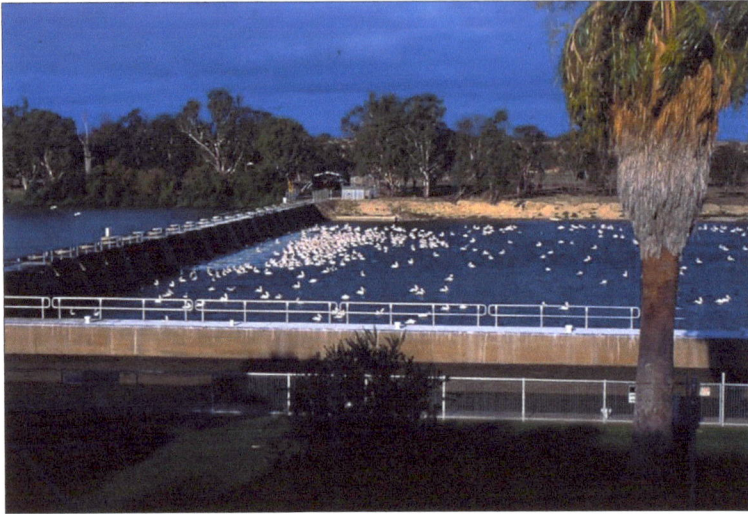

Pelicans, cormorants and egrets fishing below Weir & Lock No 1 at Blanchetown

Introduced plants and animals compete with native species for water and food, as well as changing habitat conditions. The slower flows and permanent water levels favour introduced species like European carp and European willows, giving them an advantage over native species which are adapted to highly variable flows and levels. European carp now dominate the fish biomass and out-compete native fish. European willows were originally planted to mark river channels for the paddle-steamers and to stabilize levee banks. They have spread widely through the regulated river sections which have stable water levels, dominating the river's edge. Their soft leaves, dense summer shade and lack of winter shade contrast with the scattered permanent shade and tough leaves of native river red gums. River banks dominated by European willows have less food suitable for native insects and suffer a pulse of excess nutrients and reduced oxygen in the water during willow leaf fall in autumn.

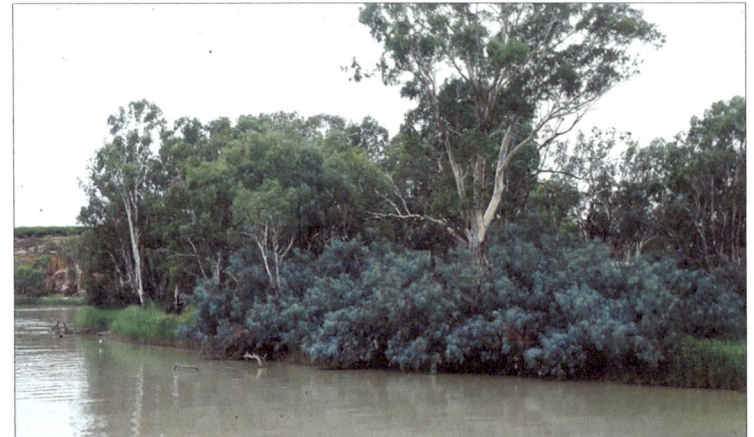

European willows (top) are displacing native willows (below) from river banks, disrupting food chains and changing habitats

The originally fresh Lower Lakes have been held at permanently higher levels by the barrages since 1940, with almost no estuarine zones of gradual change from fresh to saltwater, so the ecosystem has adapted to stable water levels and unchanging freshwater conditions. The barrages form an abrupt barrier which separates freshwater from seawater without mixing. Only 11% of the original estuary area remains, where freshwater and seawater mix. With no freshwater input to the estuary for long periods (even years) in dry times and the barrages closed for long periods, fish like mulloway can no longer migrate from Gulf St Vincent into the Lower Lakes as they once did in huge numbers.

The health of the rivers of the Murray-Darling Basin has been declining steadily. Massive extraction of water, changes to the natural pattern of flows and reduction of flooding onto floodplains have all contributed. During the Millenium Drought, the mainstream of the Murray River below Blanchetown almost dried up and the Lower Lakes fell to levels below seawater. Salinities in the Lower Lakes rose to half that of seawater and marine tubeworms encrusted every available surface, even covering the shells of freshwater turtles, leaving them unable to feed or move and vulnerable to foxes.

Since 2004, water has been recovered to return to rivers, to try to stem this decline. However, the process is being resisted and slowed by other water users. At best, the current provisions will return about <u>half</u> of the minimum needed to halt the decline in river health. This water is vital to underpin the health of all rivers in the Basin. The rivers supply the lifeblood of all Basin communities, the people and the native plants and animals, and keeping them as healthy as possible benefits everyone.

This river red gum seedling from the 2010-12 floods needs to survive to replace mature trees lost in the Millenium Drought

Black box seedlings which germinated in the 2011 flood (10-20 cm high, marked by stakes, left) received environmental water during dry years and are thriving after the 2022-23 flood (2-4 m high, right)

The Future for the Rivers of the Murray-Darling Basin

The reduced availability of water from the Murray-Darling Basin and the resulting decline in health reinforce the message that 'water is life', for everyone and everything dependent on a healthy river system: plants, animals and humans. For Murray-Darling River ecosystems, retaining enough water to sustain healthy working rivers is critical. River towns and communities, irrigation industries, tourism businesses and three-quarters of South Australia's population all rely on these rivers to sustain their water supplies, their livelihoods and their well-being.

In order to maintain healthy working rivers, we need to understand their natural history and the mysteries of how river species have adapted to the highly variable water regime of the natural Murray-Darling Basin environment. It is also essential to understand the impacts on those ecosystems of changes to river flows and patterns in water regimes since European settlement.

Hopefully the stories in this book will help to expand that understanding and to generate support for the recovery of the minimum amount of water needed to halt ongoing decline in Basin health.

The floods of 2010-12 and again in 2022-23 brought renewed life in the form of millions of seedling river red gums, black box and lignum on the floodplains, as well as young fish and waterbirds. Water recovered and returned to rivers and wetlands will help to ensure the survival of these new plants and animals, until they too can reproduce and continue the cycle of life.

Murray-Darling rivers are the only places in the world where you can see Murray cod and river red gums in their natural habitat. We need to nurture and protect this unique heritage for current and future generations.

Hope for the future -- a forest of river red gum seedlings emerged after the 2010-12 floods, with potential to replace millions of mature trees lost in the Millenium Drought, if they have access to enough water in the future

Epilogue

The ecosystems of Murray-Darling Basin rivers hold many fascinating mysteries about how they survive and thrive in such a variable and challenging environment. There have been many impacts and challenges to their survival since settlement and development of so many industries and communities which are dependent on Murray-Darling water. It is important for the future of these ecosystems and the river communities too that sufficient water is dedicated to keeping the rivers alive and functioning, so they can support the human communities that depend on them.

About the Author

Dr Anne Jensen has worked on the Murray River throughout her varied career, which has included assessing environmental impacts on river ecosystems, writing policies and strategies to protect wetlands and floodplains, coordinating on-ground projects to repair wetlands and restore their natural water regimes, and studying the processes of regeneration and recruitment in river red gums, black box and lignum so that environmental water can be used effectively.

Anne got to know river red gums, black box and lignum intimately during her PhD study, supervised by Associate Professors Keith Walker and David Paton. She spent 30 months on the floodplains of the Lower Murray Valley monitoring field sites to understand when and how these species reproduce, the conditions needed for seedlings to survive and the water requirements to maintain healthy vegetation communities on River Murray floodplains.

Anne's strong affinity with the Murray River is not surprising, as her great-great grandfathers migrated to Goolwa on the Lower Murray. One from came Scotland with Francis Cadell to participate in the paddle steamer boat trade, and the other came from Gloucestershire to become the local apothecary and Mayor of Goolwa. Her great-uncle was the captain of the last working paddle steamer PS 'Captain Sturt', used to help with construction of the Goolwa barrages in the late 1930s.

In this book, Anne has shared her love of Murray-Darling Rivers and everything she has learnt from Keith Walker and many other colleagues about river and wetland ecosystems, Illustrated with a selection of quality photographs from her extensive collection.

Useful References

Discover The Murray website: http://www.murrayriver.com.au/about-the-murray/

McKay, N & Eastburn, D (Eds) (1991). *The Murray*. Murray-Darling Basin Commission, Canberra.

Mosley, L, Ye, Q, Shepherd, S, Hemming, S & Fitzpatrick, R (Eds) (2018). *Natural History of the Coorong, Lower Lakes and Murray Mouth region (Yarluwar-Ruwe)*. Royal Society of South Australia Inc. University of Adelaide Press: Adelaide.

Murray-Darling Basin Commission (2004). *The Darling*. The Commission, Canberra.

Mussared, D (1997). *Living on Floodplains*. Murray-Darling Basin Commission and Cooperative Research Centre for Freshwater Ecology, Canberra.

Ngarrindjeri Nation Yarluwar-Ruwe Plan (Sea Country Plan)
https://8014850d-1347-477b-a48e-3d10e963e722.filesusr.com/ugd/01b606_0dbc738b1cb24f69867b58eed2c166ff.pdf

Paton, D C (2010). *At the End of the River: the Coorong and Lower Lakes*. ATF Press, Hindmarsh, South Australia.

Ponde Dreamtime story on *Discover The Murray* website: http://www.murrayriver.com.au/about-the-murray/ponde-dreamtime/

Strudwick, S (2010). *Murray River & Mallee: Australia's great River Landscape*. Murray River Tourism Pty Ltd, Norwood, South Australia.

www.ingramcontent.com/pod-product-compliance
Lightning Source LLC
Chambersburg PA
CBHW041112050426

42335CB00045B/181